Bruno Latour in Pieces

Stefanos Geroulanos and Todd Meyers, *series editors*

Bruno Latour in Pieces

An Intellectual Biography

Henning Schmidgen

Translated by Gloria Custance

FORDHAM UNIVERSITY PRESS
NEW YORK 2015

This work was originally published in German as Henning Schmidgen, *Bruno Latour zur Einführung* © 2011, Junius Verlag.

Library of Congress Cataloging-in-Publication Data is available from the publisher.

Printed in the United States of America

17 16 15 5 4 3 2 1

First edition

CONTENTS

ABBREVIATIONS FOR FREQUENTLY CITED WORKS

AT *Aramis, or the Love of Technology*. Trans. Catherine Porter. Cambridge, Mass.: Harvard University Press, 1996.

CB *La clef de Berlin et autres leçons d'un amateur de sciences*. Paris: La Découverte, 1993.

CR "Comment rédistribuer le Grand Partage?" *Revue de synthèse* 3, no. 110 (1983): 203–236.

EC "The Enlightenment Without the Critique: A Word on Michel Serres' Philosophy." In *Contemporary French Philosophy*, ed. A. Phillips Griffiths, 83–97. Cambridge: Cambridge University Press, 1987.

FR "The Force and the Reason of Experiment." In *Experimental Inquiries: Historical, Philosophical, and Social Studies of Experimentation in Science*, ed. Homer E. Le Grand, 49–80. Dordrecht: Reidel, 1990.

IC *Les idéologies de la compétence en milieu industriel à Abidjan*. Sciences Humaines, Série études industrielles 9. Abidjan: ORSTOM, 1974.

LL1 *Laboratory Life: The Social Construction of Scientific Facts*. Beverly Hills, Calif.: Sage, 1979 (with Steve Woolgar).

LL2 *Laboratory Life: The Construction of Scientific Facts*. 2nd ed., with a new postscript and index by the authors. Princeton, N.J.: Princeton University Press, 1986 (with Steve Woolgar).

LLF *La vie de laboratoire: La production des faits scientifiques*. Trans. Michel Biezunski. Paris: La Découverte, 1988 (with Steve Woolgar).

MI *Les microbes: Guerre et paix, suivi de Irréductions.* Collection Pandore 3. Paris: Métalié, 1984.

ML *The Making of Law: An Ethnography of the Conseil d'État.* Trans. Marina Brilman and Alain Pottage, and revised by the author. Cambridge: Polity, 2010.

MoE *An Inquiry into Modes of Existence: An Anthropology of the Moderns.* Trans. Catherine Porter. Cambridge, Mass.: Harvard University Press, 2013.

NB *We Have Never Been Modern.* Trans. Catherine Porter. Cambridge, Mass.: Harvard University Press, 1993.

PF *The Pasteurization of France.* Trans. John Law and Alan Sheridan. Cambridge, Mass.: Harvard University Press, 1988 (revised version of MI).

PH *Pandora's Hope: Essays on the Reality of Science Studies.* Cambridge, Mass.: Harvard University Press, 1999.

PN *Politics of Nature: How to Bring the Sciences Into Democracy.* Trans. Catherine Porter. Cambridge, Mass.: Harvard University Press, 2004.

PV *Paris ville invisible.* Paris: La Découverte/Les empêcheur de penser en rond, 1998 (with Emilie Hermant).

PVE *Paris: Invisible City.* Trans. Liz Carey-Libbrecht. Online ed., without images. http://www.bruno-latour.fr/sites/default/files/downloads/viii_paris-city-gb.pdf.

RC "On Recalling ANT." In *Actor Network Theory and After*, ed. John Law and John Hassard, 15–25. Oxford: Blackwell/Malden, Mass.: Sociological Review, 1999.

RdS "La rhétorique de la science: Pouvoir et devoir dans un article de science exacte." *Actes de la recherche en sciences sociales* 13 (1977): 81–95 (with Paolo Fabbri).

RP "Pourquoi Péguy se répète-t-il? Pourquoi est-il illisible?" in *Péguy écrivain. Colloque du centenaire, Orléans, Septembre 1973*, 76–102. Paris: Klincksieck, 1977.

RS *Reassembling the Social: An Introduction to Actor-Network-Theory.* Oxford: Oxford University Press, 2005.

SIA *Science in Action: How to Follow Scientists and Engineers Through Society.* Cambridge, Mass.: Harvard University Press, 1987.

TD "Technology Is Society Made Durable." In *A Sociology of Monsters: Essays on Power, Technology, and Domination*, ed. John Law, 103–131. London: Routledge, 1991.

TRS *Rejoicing: Or the Torments of Religious Speech*. Trans. Julie Rose. Cambridge: Polity, 2013.

VE "Les 'vues' de l'esprit: Une introduction à l'anthropologie des sciences et des techniques." *Culture technique* 14 (1985): 5–29.

ACKNOWLEDGMENTS

This book is an extended and updated version of the second edition of *Bruno Latour zur Einführung*, published with Junius Press in Hamburg in 2013 (1st ed., 2011). The themes and arguments developed in this book were first presented at the fifth European Conference of the Society for Science, Literature, and the Arts (SLSA) in Berlin in 2008.

An extended version of this presentation was published in German as "Die Materialität der Dinge? Bruno Latour und die Wissenschaftsgeschichte" in *Bruno Latours Kollektive: Kontroversen zur Entgrenzung des Sozialen*, ed. Georg Kneer, Markus Schroer, and Erhard Schüttpelz (Frankfurt am Main: Suhrkamp, 2008), 15–46. A slightly revised English version of this paper appeared as "The Materiality of Things? Bruno Latour, Charles Péguy, and the History of Science," *History of the Human Sciences* 26, no. 1 (2013): 3–28.

I would like to express my sincere thanks for the critical discussions of my initial thoughts and arguments with Hans-Jörg Rheinberger, Andrew Pickering, Michael Lynch, Robyn Smith, Steven Meyer, Didier Debaise, and Bruno Latour.

In the course of the research that followed, many people gave their kind support. Roger Guillemin, Paolo Fabbri, and Bernward Joerges agreed immediately to answer my questions. Bruno Latour provided biographical and bibliographic information and responded to my inquiries almost at the speed of light. The anonymous referees of HHS offered valuable comments and criticism.

Conversations with Bruce Clarke, Pamela Kort, Peter Weingart, and particularly my ongoing dialogue with Gustav Roßler, the principal translator of Latour's writings into German, did much to clarify my ideas. Equally important was the interaction with the students in my Latour and ANT seminars in Weimar and Cambridge, Massachusetts, and the postgraduates at the Max Planck Institute for the History of Science (Dept. III) in Berlin.

My thanks again to Skúli Sigurdsson for his helpful suggestions and pointers. I am grateful to Christian Reiss and Tomoko Mamine for reading, correcting, and commenting on parts of the German version of the book. I also thank Gloria Custance for her translation and editing work on the English text as well as Stefanos Geroulanos and Todd Meyers for their stimulating support.

Bruno Latour in Pieces

Introduction

. . . one never follows in order to reproduce.

Bruno Latour has many faces.[1] He is known to many as an ethnographer of the world of everyday technology who in meticulous studies has shown how seemingly trivial things, like a key or a safety belt, actively intervene in our behavior. Others know Latour as an essayist very well versed in theory who charged the philosophers of postmodernity—principally Lyotard and Baudrillard but also Barthes, Lacan, and Derrida—that their thinking merely revolves around artificial sign-worlds and who confronted them with the provocative assertion that "we have never been modern."

In addition, Latour is an enormously productive social scientist, who with empirical studies such as *Laboratory Life* and *Aramis* has largely contributed to defining the field of research known internationally as science and technology studies (STS). In this role, Latour is also known—despite having critiqued postmodernism himself—as an instigator of the so-called science wars, which raged especially in the United States and France over the alleged misappropriation of natural science by postmodernism-influenced representatives of the humanities and social sciences.

Over and above this, Latour has more recently distinguished himself as the visionary of a "parliament of things," in which the barriers between nature and society are to be broken down in the name of a political ecology, and he is also the programmatic spokesperson of a "new sociology for a new society," a theory and practice of social science whose core focus is the scientific and technological networks that have contributed to the genesis and dissemination of "hybrids" and "quasi-objects," of seminatural, semiartificial issues of contention—from the AIDS virus to global climate change.

Bruno Latour has indeed many faces. He is the author of an oeuvre that is as extensive as it is diverse. Since the late 1970s, Latour has produced fourteen monographs and published around 120 articles, mainly in international peer-reviewed journals. He has published a series of anthologies (including with Michel Callon and Pierre Lemonnier), several interview books (including with Michel Serres and François Ewald), and an avant-garde illustrated essay about the city of Paris with the photographer and author Emilie Hermant. He has facilitated new editions of historical science classics by Auguste Comte and Alphonse de Candolle, for example, and has contributed to the rediscovery and rerelease of works by the almost forgotten philosophers Gabriel Tarde and Étienne Souriau.

In collaboration with the artist, curator, and director of the ZKM Center for Art and Media Karlsruhe, Peter Weibel, Latour published two weighty tomes on the significant exhibitions they co-organized: *Iconoclash* and *Making Things Public*. Latour is also the author of a vast number of reviews, commentaries, prefaces, afterwords, and other short texts, which he describes as "pop articles." If one adds to all this the numerous publications spawned by actor-network theory (ANT) co-founded by Latour, then it is entirely legitimate to speak of an abundance, a plethora of Latour's and Latourian writings. His continuously updated website documents this most impressively.[2]

Against this broad background, this book provides an overview of Latour's most important works, discusses his central propositions, and places these within the context of their emergence, evolution, and diffusion.

Analogous to the two languages in which Latour expresses himself verbally and in print, the reception of his work has been concentrated above all in Francophone and Anglophone countries. Whereas in the beginning of his publishing career this was restricted to a relatively small circle of sociologists and historians of science, Latour has become one of the most frequently cited book authors in the humanities. In a ranking of the most cited authors of books for 2007 by *Times Higher Education*, Latour occupies tenth place—after Michel Foucault, Jürgen Habermas, and Judith Butler, for example, and before Ulrich Beck, Martin Heidegger, and Karl Marx.[3]

The following presentation and discussion of Latour's work considers it for the first time in its entirety: his early works on the problem of exegesis, the contributions to the sociology of science and technology, and the recent programmatic political and philosophical texts. The focus is *not* on Latour as a sociologist of science and technology, as an anthropologist, as co-founder of actor-network theory, or as political ecologist. Instead, I concentrate on the "empirical philosopher" (Latour on Latour),[4] who, beneath a top layer of changing perspectives, methods, and disciplinary contexts, repeatedly returns to *one* question: How do knowledge, time, and society relate to one another?

Latour defines knowledge as familiarity with events, places, and people seen many times over (SIA 220).[5] Building on this assumption, he asks how this knowledge is passed on within and between the various groupings and institutions of a society, regardless of whether the knowledge concerns religious beliefs, political convictions, judicial verdicts, technical skills, or scientific findings. What are the respective material bases for the handing down of experience and knowledge, and which spatial and temporal forms emerge during this process? And, finally, what effects does the passing on of knowledge have on the knowledge itself?

In this connection, *tradere*—that is, the passing on or handing down of beliefs, habits, goods, or texts—should not only be understood in the sense of historical tradition, although the passing on, diffusion, and exegesis of the Bible is a model that Latour makes repeated use of. As we shall see, this can be understood both with reference to his early interest in theological exegesis as developed by Rudolf Bultmann and with respect to an enhanced awareness of the media-historical context of printing, the Reformation, and scientific change. Latour, however, makes it quite clear that for him tradition and exegesis are eminently contemporary and current processes.

Programmatically he says, "it is my conviction that every science, including the hard ones, is defined by a certain way of practicing a peculiar kind of exegesis. Tell me how you comment on a scripture or an inscription, and I will tell you what sort of epistemology you hold on to" (EC 85–86). In the following I shall trace and elaborate the requirements and consequences of this maxim.

Latour is clearly not the first to see in tradition a crucial element of social processes. Robert Spaemann, in his studies on Louis de Bonald, has suggested that modern sociology as such originated from the spirit of traditionalism, and it is not coincidental that in this context language and nature are also very much in the foreground.[6] In addition, the coupling of the process of conveying meaning and the issue of alteration that is found in Latour's understanding of translation is by no means new. That every translator is also a traitor or—put more generally—that every act of communication is based on a misunderstanding is an idea advanced long before Latour by psychoanalysis, literary studies, and media studies.

When Latour insists that there is "no transportation without transformation" (AT 119) and that no communication takes place without "deformation" (TRS 26) or "transmutation" (RS 149), then this can be understood as an echo of Freud's famous witticism "*Traduttore—traditore!*" (To translate [is] to betray!), which that theoretician of transference and countertransference was fond of quoting, as was the polyglot historian and philosopher of science Alexandre Koyré.[7] At the same time, Latour's insistence on transmission as change can be read as a paraphrasing of an insight of the literary scholar and media theorist Marshall McLuhan: "Each form of transport not only carries, but translates and transforms, the sender, the receiver, and the message."[8]

What distinguishes Latour, however, is that he explores processes of translation and displacement that can be observed beyond and beneath purely social interactions and beyond and beneath spoken as well as written language: in the everyday handling of technological things; of maps, photographs, diagrams, and tables; and especially in the manipulation of organs, cells, and microbes that are "grown" in the laboratories of the life sciences. And it is above all in these facilities where, in the self-understanding of modern societies, an undistorted, direct, and precise access to nature and life is feasible—namely, in the splendid isolation of laboratories—that Latour identifies the operation of exegeses, of processes of imitation and alteration, the results of which reference one another as if in a network.

As a consequence, the laboratory turns into the paradigmatic space of a modernity that "has never been modern." According to Latour, the laboratory demonstrates in an exemplary way how in a literal sense the traditional is strongly anchored in the present; how far the nonsimultaneous extends into apparent simultaneity, immediacy; or, to put it another way, how closely nature and society, the given and the manufactured, the nonhuman and the human are entwined.

With this, Latour takes up a prominent position within contemporary philosophy. In the case of Michel Foucault it was the clinic and the prison that occupied a similarly central place in the social field with regard to issues of the body and the exercise of power. For Giorgio Agamben, the concentration camp is the exemplary locale from whence the state of exception that characterizes our culture of "bare life" can be comprehended. Latour, however, makes the inscription and transcription facility of the "laboratory" into the preeminent social *topos*.

The consequences are far reaching. By shifting the main focus to the laboratory, the question of socialization shifts back again from power to knowledge, from disciplining to experimenting, from discourse formations of grand dimensions to local associations of material and semiotic components.

In the following I shall demonstrate that this literal change of terrain is on the one hand the result of empirical studies conducted in the sociology of science and technology, which engaged in a very concrete way with the laboratory. On the other hand, the shift mentioned reveals a specific closeness to the philosophy of Gilles Deleuze and Félix Guattari.

Latour has repeatedly cited Michel Serres to elucidate the central motifs of his work, for example, the subject of translation or the alleged continuity between early and late modernity. The philosophy of Jean-François Lyotard has also been an important stimulus for Latour's work. As we will see, this applies especially to Lyotard's project, following Nietzsche, to develop a "third" sophistic art of simultaneously arguing a topic from both sides, the *dissoi logoi*. Latour's concept of a "symmetrical anthropology," which treats the representation of humans and nonhumans together, is strongly committed to this approach. In addition Lyotard's thoughts on the meaning of rhetoric and performance in modern science find some resonance in Latour (and vice versa),[9] even though he emphatically dissociates himself from Lyotard's postmodernism in *We Have Never Been Modern.*

At an earlier point in time, however, and with more lasting effects, it was the references to Deleuze (and Guattari) that were defining for his work. An indication of this is found in an interview from the 1990s in which Latour declares, "Deleuze is the greatest French philosopher (along with Serres). . . . I have read Deleuze very carefully and have been more influenced by his work than by Foucault or Lyotard."[10] In the same interview, Latour says that the concept of the "rhizome" as used by Deleuze and Guattari is the apposite name for what he himself calls a "network." And, in fact, this famous term from *A Thousand Plateaus*, which unlike the structuralist view of language as a branching structure directs attention toward the local assemblages and multiplicities—"stem tubers"—that consist of very different material and semiotic components,[11] is found increasingly in Latour's more recent works (e.g., PN 51).

The notion of the network, though, is but one of the most overt connections between the philosophical terms created by Deleuze and Guattari and Latour's theoretical conceptions. As I will show, it is the question of tradition itself—understood as the interplay between "difference and repetition"—that represents a bridge between the two enterprises. That repetition is a machine to produce differences is one of the Deleuzian insights put forward by the early Latour (Chapter 1). That the confrontation of knowledge and power, reason and strength, cognition and politics can be overcome through recourse to Nietzsche's concept of "force" is another (Chapter 5). Even the famous "Follow the actors!", the basic principle of the anthropology of science as conceived by Latour, derives from a concept of Deleuze and Guattari's: namely, the "nomad, ambulatory sciences." According to the authors of *A Thousand Plateaus*, the approach of these sciences "consists in following a flow in a vectorial field" in order to respond to the singularities of the material: "Don't go for the root, follow the canal."[12]

Further connections to Deleuze and Guattari become manifest via authors to whom Latour refers approvingly, as do Deleuze and Guattari, like the philosopher of technology Gilbert Simondon, the social philosopher Gabriel Tarde, and the writer Charles Péguy.[13] However, the connections do not just rely on shared references to authors, concepts, and topics. They are also a matter of form. The famous couple "Deleuze and Guattari" relied on a collaboration between philosopher and psychotherapist, that is, between theoretician and practitioner. This presages the dialogue between reflection and empiricism that still continues today in Latour's work. It is the heterodox practice of a research endeavor that constantly moves from systematic investigation of an ethnographic, sociological, or historical field to conceptual elaboration, from there moving back to empiricism, and afterward returning to theory, which is characteristic for his work. And this oscillating movement from exploration to elaboration is responsible for the many faces of Latour: from ethnographer to essayist, from science scholar to politician and philosopher.

However, the connections to Deleuze and Guattari are not seamless. Latour does not share the philosopher-therapist duo's orientation on historical materialism nor their interest in identifying movements of sociotechnological subversion. To him, the main focus is on establishing a new kind of cartography, one that can be used to trace phenomena and processes of tradition. As a result, any revolutionary impulses recede firmly into the background.

To substantiate these arguments, this book traces the genesis and development of Latour's oeuvre from the early publications on the problem of repetition in the work of Péguy and the issue of an ideology of competence in Abidjan (Chapter 1) to the semiotics and sociology of the everyday at the Salk Institute (Chapters 2 and 3); from the historical-philosophical analysis of modernization, which is associated with the name of Louis Pasteur (Chapter 4), to the general view of "technoscience" that is provided in *Science in Action* (Chapter 6); from the first rudiments of actor-network theory, which Latour developed in the context of his philosophy of "irreductions," and the project for a history of science as "a history of things" (Chapter 5); to the intensified focus on sociology and philosophy of technology (Chapter 7) and the recently envisaged orientation on a new "ontology," which propagates a pluralism of "modes of existence" (Chapter 8).

Given the sheer mass and the ongoing expansion of the Latourian oeuvre—he himself describes it as a "moving target"[14]—it is obvious that this book sets certain areas of focus and has clear limits. For example, the many debates and polemics, which were triggered by Latour himself or his colleagues and critics, are not presented and discussed *in extenso*. The "science wars" would deserve their own book; they can only be briefly touched upon here (Chapter 7). By contrast, the early writings of Latour, that is, before the publication of *Laboratory Life* (1979), will be an important focus (Chapter 1).

By the side of the road that led Latour from studies in Dijon via sojourns in Abidjan and La Jolla to Paris, some comparatively unknown figures will emerge, like the philosopher and Bultmann specialist André Malet, the ethnographer and Côte d'Ivoire expert Marc Augé, and the semioticians Paolo Fabbri and Françoise Bastide. Yet only when the stimuli for the development of the Latourian oeuvre that originate from these figures are accounted for is it possible to assess more precisely the significance of this oeuvre. To this, my book hopes to make a contribution.

ONE

Exegesis and Ethnology

Beaune is one of France's most famous and important wine centers. The small city in Burgundy is also the birthplace of two important scientists: in 1746 the mathematician Gaspard Monge, and in 1830 the physiologist Étienne-Jules Marey. Unsurprisingly for this region of the world, wine is one of the connecting links between Monge and Marey. Both scientists came from families of winegrowers and wine merchants—the two families had actually joined forces for a time in the late eighteenth century. And even the scientific work of these two sons of Beaune was associated: although their subjects could not have been more different, Monge and Marey shared a fascination with lines.

In the early nineteenth century, Monge became one of the founding fathers of descriptive geometry. Already at school he had made a large-scale plan of his home town using drawing instruments he constructed himself. In the 1860s Marey made a name for himself as a laboratory scientist specializing in physiology and as a proponent of all instruments and techniques for graphically recording vital functions in the widest sense—from actions of the heart and lungs to speech and the flow of water. In many books and articles Marey praised the virtues of the *graphic method*, which he said enabled him to get the phenomena of life to speak "in their own language"—or, even better: to draw—in the laboratory.[1]

Latour was also born in Beaune into a family of wine growers, and he also has "a passion for the trace." In his sociological studies of science, for example, he frequently talks about the *inscription devices* that are used in modern laboratories to record and collect data and visualize it in clearly arranged curves. In a countermove, as it were, in his sociological studies of technology he develops a special graphic method in which "sociotechnical graphs" are used to map how human beings interact with technical beings. And when Latour investigates the visibility of Paris as an ethnographer, he takes along a (chrono)photographer, who translates the different sequences of their exploration into curved series of images (PV 21–22).[2]

It is tempting to see Monge, Marey, and Latour's partiality for lines, curves, and graphs as connected to their native Burgundian landscape of wine-growing countryside. For isn't it possible to see the training of vines, by attaching them to stakes, trellises, or pergolas, as an interesting conjunction of lines? On the one side the flexible lines of the climbing and winding plants and on the other the rigid lines of the vine training system? Peter Sloterdijk, in a tribute speech honoring Latour, went so far as to speak of the "primary Burgundism" of Latourian thought. According to Sloterdijk, the dense fabric of grapevines, terroirs, rhizobacteria, fermentation processes, the vintner's art, glass industry, and haulage firms, which is so typical for the Beaune context, represent the basic model for Latour's recurrent theme that natures, humans, things, and technologies belong together.[3]

It is indeed possible to understand wine as an exemplary object of Latour's empirical philosophy, a "quasi-object" that is situated equally between nature and culture, as well as between religion and science, on the one side, and matter and signs on the other. For wine is first and foremost an organic substance that has been made for thousands of years and has developed into a fine art and a science (including through pasteurization, which Maison Louis Latour used early on for its red wines).[4] Wine is also a cultural symbol that in all ages past as well as today possesses a fascinating aura and an enormous power of attraction. And, after all, it is the enjoyment of tasting wine—"Let us suppose that a cellar in Burgundy invites you to a wine tasting . . ." (PN 84)—that Latour judges to be a fitting example of how the interplay of sensations and inscriptions, of organs and instruments leads to an enhancement of the ability to discover in an ostensibly well-known reality an ever greater number of experienceable differences and variants.

Studies in Dijon

In the mid-1960s Latour started to detach himself from his home town. In nearby Dijon he began to study philosophy in 1966. Dijon is also not just any town. In the 1910s, Lucien Febvre taught at the local university. As is well known, Febvre included disciplines such as linguistics and ethnology in his historical studies as important resources and in this way contributed decisively to establishing the Annales school of historical thinking. The same disciplines, in particular linguistics and ethnology, would be of particular interest to Latour in his writings on the sociology and history of the sciences. Furthermore, at the University of Dijon one of the founders of historical epistemology, the most influential form of Francophone history and philosophy of science, worked for ten years: Gaston Bachelard. From 1930 to 1940 Bachelard taught and researched in Dijon, which is where several of his best-known works were written, including *The New Scientific Spirit* and *The Psychoanalysis of Fire* as well as *L'intuition de l'instant*, a hitherto untranslated tribute to Bachelard's friend and university colleague, the regional historian and writer Gaston Roupnel.[5]

Latour has no time for Roupnel's ecomysticism, and even to Bachelard he refers only for a limited period of time. In the late 1970s Latour uses Bachelard's concept of "phenomenotechnique" when it is a matter of describing the construction of research objects in the scientific practice of laboratories. Later, Latour will increasingly distance himself from the "French style" of the history of science, which, following Bachelard, was developed in Paris under the aegis of Georges Canguilhem in particular. In this period, the mere mention of the words "epistemology" or "epistemological break" suffices to provoke reactions of dissociation in Latour.[6]

Conditions in Dijon were accommodating to the distancing of oneself. Whereas in the French capital enthusiasm was growing for both the student movement and historical epistemology—especially in Pierre Macherey, Dominique Lecourt, and other students of Canguilhem and Louis Althusser—in Dijon the philosopher Jean Brun was the dominant figure. Brun was a reputed expert on the ancient Greek philosophers and a gifted educator who, following Kierkegaard, developed an existentialist view of the human subject, inwardly divided but seeking balance. Despite that the early Canguilhem was one of his teachers when he was at school, Brun's interest in the problem of technology was not motivated by epistemology but rather by the history of philosophy. Thus it does not appear that this interest of Brun's led him to offer any courses in the philosophy of science or indeed in the sociology of science.[7]

Another young philosopher, who was teaching in Dijon at this time and who had considerable influence on Latour, maintained an even greater distance to Bachelard: André Malet. After completing his Ph.D. in Catholic theology, in 1957 Malet converted to Protestantism and began a second dissertation, this time in philosophy. Titled *Mythos et logos*, it was an in-depth analysis of the work of the German Lutheran theologian and biblical scholar Rudolf Bultmann. Malet's supervisor for the dissertation was the philosopher Paul Ricœur. This may be one of the reasons why the initial question posed in Malet's thesis, published in 1962, does not sound much like theology: "What is objectivity?" The orientation becomes more comprehensible in light of the central theme of the dissertation, which is Bultmann's project of "demythologization" (*Entmythologisierung*), that is, examination of the biblical texts with regard to their authentic content, their existential core. Malet's thesis concludes with a discussion of the relationship of Bultmann's theology to the philosophy of Martin Heidegger titled "Theology and Ontology."[8]

In the supplementary dissertation, which at that time was required, Malet also engaged with the beginnings of modern biblical exegesis, particularly Spinoza's *Tractatus Theologico-Politicus*. When the supplementary dissertation was published in 1966, Malet had already moved from Paris to Dijon. There he became Brun's assistant, and one of his first projects was to undertake the translation of Bultmann's *History of the Synoptic Tradition* into French—a work that Latour still cited in the 1980s in the preliminary studies for his "anthropology of science" (CR 232).[9]

The young student was seriously impressed by this former Catholic priest who had a consuming interest in Bultmann *and* Heidegger. As Latour recalls, he "spent" four whole years with Malet, and during this time he studied Bultmann's work "intensively."[10] One encounters the traces of these studies in his oeuvre time and again. In 1975 Latour completed his studies with a Ph.D. thesis, the title of which—*Exegèse et ontologie à propos de la resurrection* (Exegesis and ontology, with reference to the resurrection)—can be read as a reference to Bultmann. Four years later, in *Laboratory Life*, he cites the "form criticism" co-developed by Bultmann, when it is a question of the "existential interpretation" of scientific reports and accounts (LL1 169), and in 1984, in the first pages of his book about Pasteur, he invokes the sections of Spinoza's *Tractatus Theologico-Politicus* that deal with the reading of the Scriptures to make a case for a new kind of "exegesis" of scientific texts (PF 7). During an interview in the mid-1990s Latour even made a point of stressing that "I was trained in philosophy and biblical exegesis," and as he has recently stated more precisely, it was without doubt his study of Bultmann that directed his attention to the reading and writing aspects of scientific practices.[11]

However, the road from Bultmann's theology to the sociology of science and technology is not a direct one, and Malet's notions of "incarnation *in actu*" and "tradition (in an active sense)," as heavily as they are inspired by contrasting the ancient Greek world of science to the biblical world of religion,[12] cannot be readily transformed into a notion of *Science in Action*.

Péguy's Inscriptions

Initially, Latour even took up the question of exegesis in a completely different way. This is demonstrated by his very first publication, a lecture held in 1973 but not published until 1977, which was on the theme of repetition in the work of the writer Charles Péguy.

Born in 1873 and killed in action in 1914, Péguy was one of the most important philosophers and poets of French modernism. He wrote plays, dialogues, religious epic poems, essays, and polemics, including against Émile Durkheim's sociology, which at that time was in the process of establishing itself as a formal academic discipline. Péguy also founded, edited, and published the journal *Cahiers de la Quinzaine*, in which, between 1900 and 1914, appeared contributions by authors such as Henri Bergson, Anatole France, Jean Jaurès, and Romain Rolland as well as his own essays. In his text Latour focuses on one of Péguy's most famous works, *Clio: Dialogue de l'histoire et de l'âme païenne* [Clio: Dialogue between history and the pagan soul]. He engages with both versions of this dialogue published posthumously, *Clio I* and *Clio II*, and in the context of a "structural reading" compares them to the synoptic gospels of the New Testament.

It is worthwhile to take a closer look at Latour's unconventional reading of Péguy's works, because the text already lays out his central theme: namely, the problem of the social tradition of experience and knowledge. Further, this early publication exhibits typical stylistic features that we will also encounter in later works, such as the explicit reflections on a role that is merely played by the author and the utilization of graphic schemata to illustrate what has been written or said. In addition, the essay on Péguy—via the Malet-Bultmann context, so to speak—conveys the encounter between Latour and Deleuze, which we shall look at more closely in the following chapters.

As a first step *Clio* can be seen as the result of transposing Bergson's philosophy of time onto the problem of history during Péguy's gradual turning to Catholicism. Similar to the way in which Bergson contrasts time and duration, intellect and intuition, Péguy differentiates between history and tradition, science and experience. To illustrate this difference, for example, in *Clio* history is compared to a long railway line that runs along the coast and that allows one to stop at any station one wishes. In this metaphor tradition—collective memory—appears as the coast, with its marshes, people, fishes, estuaries of rivers and streams, as life on land and life in the sea.[13]

Clio develops this two-part metaphor within the context of a critique of the modern era and the great importance it attaches to the systematic study of history. That research methods and instruments are deemed to have primary importance in the way the discipline of history operates, Péguy considers one of the greatest errors of modern times. As he writes elsewhere: "As though ignorance of the present were a sine qua non for access to knowledge of the past."[14] The countermodel of history that Péguy proposes against the systematic endeavor of the Sorbonnards is oriented on secondary-school teachers in rural areas, who act upon the basis of their "tacit experience" as (family) people and (world) citizens, that is, as committed witnesses of history who, caught in a precarious social situation, try to attain orientation about and within history.

Against the methods of professional historians, which are inspired by a belief in progress and therefore only ostensibly neutral, *Clio* evokes an ethos of memory, preservation, and self-reflection. With concepts such as *rendre*, *entendre*, *interpréter*, and *représenter* Péguy outlines a process of remembering and narrating "which is in the service of memories of the bygone world that will be passed on."[15]

Péguy assumes that the basic form of this type of handed-down memory is expressed in the same kind of language of signs, forms, and figures in which the Christian God addresses humankind. For him the fullest expression of this language is the fact of the incarnation—similar to Malet's notion of tradition *in actu*. Accordingly, the muse of modern historiography seems highly problematic, for she merely appears as the "young lady of the records." The author of *Clio* can only see in her a pillar saint of a religion of "stamped pieces of paper": "Always this superstition, this cult of the inscription. Always this idea that inscription leads to the action, is the action."[16]

Péguy confronts this cult with a more authentic world of inscription. Superficially it manifests itself in church architecture. With these buildings in mind Péguy brings together script and body, prayer and stone:

> Works that do not detach themselves one iota from the cult and prayer and worship; so little that they are like, they literally *are* a physical inscription, a temporal inscription, an inscription carved in stone, stone, taking place in stone, of the service and prayer, the innermost, and of the most fervent worship. The worshipped body, works of stone, inscriptions, incorporations.[17]

Here inscriptions are not merely a matter of pen and paper; rather, they are embedded in a space where encounters between people and things take place; they are not a silent archive but a physical and meaningful ritual in which bread and wine play decisive roles.

The Problem of Repetition

Latour considers Péguy as "probably the greatest French prose writer and no doubt the deepest philosopher of time,"[18] and for him *Clio* is "probably the most profound study on the articulations of the various historical and religious times" (PF 258). Latour's early study of this work was certainly marked by the context of biblical exegesis sketched above. At the same time, however, it was also determined by the fact that Latour began to engage with the philosophy of Gilles Deleuze.

In 1968 Deleuze refers to Péguy in his major work *Difference and Repetition*, pointing out that Péguy's philosophy of history involves a specific conception of repetition and the event. Similar as in Kierkegaard, in Péguy repetition is superseded by the notion of a going back to the past that is always the same, to then be redefined as a forward-looking, open process, which "recollects forward" (as Kierkegaard puts it) about something. According to Deleuze, Péguy does not conceive of historical tradition from the perspective of the endpoint of a series of events; he proceeds on the basis of the unique event, which is the reason for the process of passing something on in the first place: it sets the process in motion and continues it.[19]

To illustrate this Deleuze cites two passages from *Clio*. The first concerns Claude Monet's paintings of water lilies, *Nymphéas*. Péguy's muse of history says about this series of paintings:

> Which of these twenty-seven and of these thirty-five water lilies is painted the best? The most logical thing would be to say: The last one, because he was at the height of his skills. But I say: On the contrary, it was actually the first, because then he was least skilled at doing it.[20]

In this way repetition, as Deleuze says, "interiorizes . . . itself"[21]; that is, repetition is defined as a relationship and behavior toward an event that is not similar or equivalent to anything else. The singularity of the first painting of water lilies triggered a creative process in Monet that was both repetitive and differentiating; it continued and led to further paintings of water lilies.

That Péguy also argues in this manner with respect to historical events is illustrated by the second example Deleuze cites from *Clio*. Péguy suggests that, eventually, the storming of the Bastille cannot be attributed to the logic of history or external reasons or interests but was due to the basic spontaneity of the people:

> The Bastille had never done anything to them [the French people]. The storming of the Bastille . . . was actually a party, it was the first celebration, the first commemorative ceremony and the first anniversary, so to speak, of the storming of the Bastille.[22]

Just as Monet's first painting of water lilies brought forth all the others that followed, it is the historical event itself, as Deleuze says, "which celebrates and repeats in advance all the Federation days."[23] As a unique happening in all subsequent anniversaries it recollects itself "*vorlings*," to paraphrase Kierkegaard.

In *Difference and Repetition* Deleuze moves Péguy to stand beside Kierkegaard and Nietzsche in a "triptych of priest, Antichrist, and Catholic," which stands at the beginning of his endeavor to make repetition the fundamental category of a new philosophy.[24] Yet it was not only Péguy's *theory* of the (historical) event that made him one of the "great repeaters of history"[25] for Deleuze. Péguy had also developed a new means of expression, a new *practice* of writing with which "to put metaphysics in motion, in action."[26]

What Deleuze is referring to is, on the one hand, the theatrical and dialogic aspect of Péguy's works and, on the other, his characteristic style that flows and is interrupted, which has been subjected to analysis since the 1920s by Leo Spitzer and other linguists. According to Deleuze Péguy uses series of synonyms and contiguities to generate small differences of meaning that open up the "internal space" of the words.[27] Thus for Péguy repetition is not only a theme; it also finds expression performatively, so to speak. It is precisely this observation that marks the point at which Latour's exploration of Péguy's work begins.

Exegeses, Rereadings, Revisions

Accordingly, his opening questions are: "Why does Péguy repeat himself?" and "Is Péguy unreadable?" To find answers Latour approaches *Clio* not via the actual text but via the reader. On the reader Péguy's repetitions—"these incessant digressions, these monstrous paragraphs, these violent accelerations" (RP 79)—exert a specific, precisely calculated effect, namely, to override all the usual criteria of readability. In this sense Péguy is indeed "unreadable" because the fundamental aim of this writer is to deter readers from pursuing their usual habits of reading, comprehending, thinking, and even living. Expressed in general terms, "Repetition is the war machine devised by Péguy aimed at combating the refrain [*ritournelle*] and saying the same thing [*rabachage*]" (RP 80).

In concrete terms *Clio* is first and foremost a "manual for how to read correctly." In contrast to the modes of reading that are entertaining, historical, or clerical, this dialogue offers a way of reading that "breaks with old habits" and allows the originality of a text to reappear. Proper reading makes a text start over again and renders it an event, which approaches the reader in the present moment from far away.

According to Latour, one of the characteristics of this kind of reading is a change in the usual direction. Reading does not any more proceed horizontally but is turned by ninety degrees and proceeds vertically. It is no longer a matter of the superficial sequences of words in lines, nor is it about the lateral connection of one text to others; instead, it is a connecting back to the depths of time of what has been read, the recourse of what is being read in the present to the past event of writing. Péguy's distinction between history and tradition, science and experience, is found here at the level of the reader and the reader's willingness, or rather responsibility, to read an old text in a truly new way.

In a second step Latour undertakes a vertical reading of *Clio* in a literal sense in which he engages with both versions of the dialogue. In the course of this his quasi-ethnological distance to the text becomes a real ethnological distance: in this section of his study Latour adopts a method suggested by Claude Lévi-Strauss to investigate myths in which text units are arranged in a table and subdivided according to their "rhythm" and "tonality." On the basis of this "grid of reading" Latour claims to be able to infer the "precise logic" of *Clio I* and *Clio II*, which he then presents in the form of a schematic drawing. Like the table that divides up the text of *Clio*, this diagram facilitates a calculated distancing of Latour from the literary text, which according to him is an important mark of any "correct exegesis." Latour assumes that a commentary on a Péguy text is only apposite if it does *not* utilize any of the words that appear in the text "because otherwise it would rehearse the enigmatic text *in an ordinary sense*" (RP 85). Thus no reproduction and no citation is aspired to, only differentiating repetition, in Péguy's sense of the term. As a consequence, the distance that had been attained previously changes quietly into proximity, into cooperation.

The third part of Latour's reading of *Clio* focuses on Péguy's often-quoted Catholicism. Latour argues that the Catholic orientation of this writer cannot be inferred from his use of religious terms but instead primarily from his way of writing, his style. Péguy does not (or does not only) adapt, popularize, or modify the substance and tenets of the Catholic religion; rather, he allows them to begin again as new: "By means of Péguy's repetition each person hears the word of God in their own language; *repetition renews the work of Pentecost*" (RP 93). Therefore, Péguy is not a prophet of any variant of religion. For Latour he is nothing less than a new evangelist. Péguy's repetition repeats the repetition of the three synoptic gospels (Matthew, Mark, and Luke): "He [Péguy] is really the one who once again and using the same method brings the Glad Tidings of the past withdrawal, the Glad Tidings of past events, of the perennial openness of the event" (RP 94). In Latour's opinion, if the gospels were read vertically after the manner of Lévi-Strauss's method, this would actually reveal that they have the same internal structure, the same rhythm as *Clio*. In this way the gospels and their tradition are relocated against the backdrop of reflection that Péguy sets up in the *Clio* dialogue: "the great and constant movement of exegeses, rereadings, and revisions of the Holy Scriptures belong exactly to the machinery that *Clio* has made its subject" (RP 93).

Ideology

In the introduction to this first publication Latour introduces himself as an *agrégé* of philosophy who is the top student of his year (1972) and who, despite his "youth," wishes to make a contribution to the hundredth anniversary of Péguy's birth. This is reminiscent of feigned authorship à la Kierkegaard when he quotes from a letter sent by "Bruno Latour" in which the author requests that the organizers of the event agree to his participation (RP 77–78). The actual lecture, the "Communication," follows after this prelude.

The next role that Latour assumes is a very different one. National service wrenched the young philosopher out of the relative tranquility of Dijon and plunged him into the teeming bustle of the African city of Abidjan, with its over a million inhabitants. The Republic of Côte d'Ivoire (Ivory Coast) gained its independence in 1960; Abidjan is the economic center and until 1983 was the official capital city of the country. For two years, from 1973 to 1975, Latour worked there for the Office de la Recherche Scientifique et Technique Outre-Mer (ORSTOM), a development aid agency.

Based on his lecture on Péguy, Latour began to work on his Ph.D. thesis while at the same time conducting a survey for ORSTOM in Abidjan on the "ideology of competence" in industrial circles. He was supported by his colleague Amina Shabou, who later became a medical sociologist. The subject of the study was a precise issue in development sociology: namely, which factors are responsible for the fact that managerial positions are not filled with Ivoirians but with people from France or other Western European countries. The working hypothesis is that prevailing assumptions and prejudices as to the incompetence of Ivoirians are the reason.

With regard to methodology Latour and Shabou modeled their study on qualitative social research. They conducted around 130 semistructured interviews with Africans and Europeans at various levels of the economic hierarchy—from directors and managers to master craftsmen and foremen to manual workers and apprentices and trainees. The evaluation of the interviews was based on content analysis, which focused on identifying keywords, basic motives, and typical forms of argumentation.

The study conducted by Latour and Shabou took place in an intellectual environment that had been influenced by the Ivory Coast expert Marc Augé. Since 1965 Augé had undertaken ethnological field trips in Ivory Coast, focusing particularly on the Alladian peoples, at that time numbering around ten thousand persons, who lived on the edge of a large lagoon to the west of Abidjan. Until 1970 Augé was the director of ORSTOM in Abidjan, after which he took up a position at the École des Hautes Études en Sciences Sociales in Paris. There, from the 1980s, he focused increasingly on ethnological research of his urban environment. Internationally Augé is particularly known for his publications resulting from this later period of his career, for example *Non-Places* and *In the Metro*.[28]

In spite of the obvious contrast between the local contexts of Dijon and Abidjan certain connections do exist between Latour's philosophical work on Péguy and the development sociology study on the ideology of competence. The most obvious one is attached to the term "ideology" which figures prominently in the study's title: "Les idéologies de la compétence en milieu industriel à Abidjan." Basically, this is just a new name for religion, beliefs, and traditions. In the early 1970s ideology became a fashionable buzzword, which was used especially by Marxists. At the same time, the term was often referred to in French structuralism and psychoanalysis. In his ethnological investigations Augé followed this general trend. Based upon his case studies of superstition, magic, and faith healing among the Alladian people, he argued that "organization" and "representation" of a society are always found together. As a consequence he underscored that ideology is not a rigid and closed system that is imposed on the "actual" social reality.

According to Augé, ideology is not a totalitarian system from which there is no escape and that gives rise to "everything," society as well as its organization and representation. In fact ideology is inherent in society as a system of "semantic production," hence of "constructing the world." It may have its own elements and special rules, but it is open to being used by the social subjects in a great variety of ways. Therefore, ideology is a largely independent "ideo-logic," and the ethnography of this ideo-logic leads to a "sociology of *médiations*." By this Augé means sociology that for the overall logic of a social system is "not reducible [*irréductible*]" to the logic of one of its parts.[29]

The Production of Lack

Latour's work in development sociology fits in with this conception up to the point when Latour speaks affirmatively about what Augé had deprecatingly referred to as the "meta-anthropology" that Deleuze and Guattari developed in their *Anti-Oedipus*.[30] In the beginning it seems that Latour's study is indeed an interrogation of "ideo-logic" as understood by Augé. His analysis deals with the various "discourses" concerning the problem of competence and incompetence. These discourses, explains Latour, are characterized by heterogeneity because they are based on a "network" (*réseau*) of protagonists and because they "circulate" in rather different places: factories, offices, cafés, and schools. A statistically representative survey on the question of the ideology of competence seems not feasible; however, it is possible to produce a "qualitative analysis" (IC 4).

Thus the ideology of competence was not interrogated as to its truth or validity but investigated as a rather independent phenomenon, as an authentic report about "pseudo-facts" (IC 5). According to Latour, the discourses in question are mainly of a racist nature, on the part of the European and the Africans alike. Besides mystifications, there are frequently "pseudo-scientific" elements included in these discourses. On the whole these are tenacious beliefs that, to put it in Augé's terms, determine the organization as well as the representation of a society but that only in exceptional cases challenge its basic structure.

Latour goes one step further, however, when he becomes interested in what he calls the "creation of incompetence" and more generally "the production of lack." Here he uses an expression from *Anti-Oedipus*, which deploys Marxist arguments against the idea put forward by Lacanian psychoanalysis that "lack" is constitutive for the human being as such.[31] Analogously Latour declines to explain the genesis of ideologies with psychological theories. In his view, it has to be ascertained how the problem thematized by the ideology—incompetence—is created in concrete settings, for example, in the factory or in schools.

A visit to a vocational training school and the interviews conducted there soon lead to a diagnosis: "In fact the trainees did lack a thoroughgoing, direct, and primary relationship to the machines" (IC 57). This hiatus between theory and practice is the "epistemological obstacle" in the vocational education of the Ivoirians, who would judge themselves and be judged by others to be competent.

With Bultmann, here one could speak of an "existential interpretation." Ideology is referred back to a certain form of existence (*Dasein*), while the ideal constitution of this existence is oriented on the machine world of *Anti-Oedipus*. Following Deleuze and Guattari Latour argues for a largely self-determined interaction with technical objects as a basis for an open society. Ivan Illich's advocacy of "convivial" dealings with technology, which Deleuze and Guattari had cited approvingly, is also mentioned by Latour in this context (IC 75).[32] Thus once again the world of representation is contrasted with the world of production.

Latour goes even further. He places the "production of lack" observed in Abidjan as a whole in the perspective of a "deterritorialization" in the sense of Deleuze and Guattari. This concept is used in *Anti-Oedipus* to describe the specifically capitalist tendency of uprooting workers. Latour refers to the respective passages in Deleuze and Guattari's famous book and adds with regard to the situation in Abidjan:

> The relationships between two individuals from the same region, the same location, the same ethnic group become increasingly weaker than the relations between any individual and institutions such as the state and the two universal equivalents: money and knowledge. The entirety of these lines of flight charts the surface area of capital. (IC 75)

Here, at the very latest, the difference between Latour and Augé becomes obvious. Whereas the latter emphasizes the independence of ideology as a genuinely social phenomenon, Latour turns his attention to the base, to capital. At the same time a motif comes into view that will prove important for Latour's later work. It is the project to analyze individual institutions in order to show how they produce ideologies and how they consolidate them:

> How does this factory or this school actually function if one examines the circulation of information, of power, and of money? Who possesses which document? Who is able to do what? Which part is mastered by everyone, and which parts remain in the dark for them? (IC 67)

One could say that at this point the vertical reading of Péguy gets turned by ninety degrees. The transmission of events is no longer understood as a process that takes place within time but as a course of events within the space of an institution. Life in such spaces is characterized by a specific form of exegesis, by "circulation of information" and the "possession of documents." From this perspective it becomes possible to comprehend not only factories and schools but also laboratories as places of the transmission and exegesis of written documents.

TWO

A Philosopher in the Laboratory

Latour's next stop was the Salk Institute for Biological Studies in La Jolla, California. From October 1975 to August 1977, as a participating observer he collected the data on which *Laboratory Life*, the book he wrote with Steve Woolgar, was based—a pioneering study in the anthropology of science.

The Salk Institute, which is located directly by the sea in an area of outstanding natural beauty near San Diego, was founded by the clinical medicine researcher and virologist Jonas Salk. In 1955 Salk had presented the first effective vaccine against paralytic poliomyelitis. Although he had drawn on the work of others in developing his vaccine—the majority were later honored in the "Polio Hall of Fame"—it was Salk whom the U.S. mass media singled out as the man who vanquished this horrific disease. In the following years Salk used his charisma and his enormous popularity to gain the support of foundations, charities, and other sponsors for an ambitious plan. He wanted to create an independent research center where a community of the world's best scholars interested in different aspects of biology and medicine could come together to conduct creative research.

In addition to the funds and technical equipment required for the institute, Salk regarded the buildings of such a center, their internal structure as well as their external location, as a very important if not decisive factor. In 1959 he began to work with the architect Louis I. Kahn to develop an appropriate style of architecture. Kahn had risen to prominence with his futuristic new main building for the Yale University Art Gallery, and in the late 1950s his plans for the Richards Medical Research Laboratories building at the University of Pennsylvania achieved international acclaim. This design featured three towers containing laboratories connected with a central service tower; the laboratories used by people, the "served spaces," were separated from the "servant spaces," which contained mechanical systems, lifts, animal quarters, etc. Designed originally for each floor to be one large room, partitions could be employed to accommodate the changing needs of the scientists. Salk had similarly open structures in mind for the research center he was planning.

Shortly after he and Kahn commenced their collaboration, Salk got the go-ahead to build on a site in an exceptional location—a mesa in La Jolla. In June 1960 in a special referendum the citizens of San Diego voted overwhelmingly to give the land to Salk for his institute. Also nearby was the new University of California campus planned for San Diego. Situated on cliffs offering breathtaking views of the Pacific Ocean, Kahn's building complex for the Salk Institute for Biological Studies—"one of the major works of twentieth century architecture"—took shape over the next few years.[1]

The complex features two symmetrical building structures facing each other lengthways; they are separated from and connected to each other by a grand courtyard that gives onto the ocean like the stage of an open air theater—an impression reinforced by a stream of water that flows down the middle of the courtyard in the direction of the ocean. The deep embedding of the building complex into the landscape is further emphasized by the fact that from the courtyard the laboratories are hardly visible. The actual rooms where research is done are beneath the courtyard and in the rear sections of the two symmetrical buildings. At the front are the scientists' studies, which look out onto both the courtyard and the ocean.

Kahn's client referred to these studies as "monk's cells," which points to scientists' need for concentrated study and also indicates that Salk saw his institute in some measure as a spiritual place: research as a creative act in a welcoming and inspiring environment that provides both a retreat and the necessary comfort for researchers to investigate and explore. With the clean lines of its design and its durable, basic, and low-maintenance materials (concrete, wood, glass, and steel) it is not difficult to see this institute as a modern monastery—or as a high-tech version of the Grand Canyon, where the simultaneity of silent depths and openness has a similarly awe-inspiring effect on the beholder.

Completed in 1965, the buildings were soon in use. It did not take Salk very long to attract a group of eminent medical scientists and biologists to his institute: Jacob Bronowski, Leslie Orgel, Leó Szilárd, and the later Nobel laureate Robert Holley, as well as luminaries like Francis Crick, Jacques Monod, and Warren Weaver as nonresident fellows. Despite or perhaps because of the high profile of its staff, the Salk Institute was not eager to focus its activities exclusively on the natural sciences. From the start, bridges were built to literature, philosophy, linguistics, and the social sciences. Jacob Bronowski in particular personified this tendency. Alongside his research work in mathematical biology, he published studies on the history of literature and the history of science, was interested in the subject of imagination in art and the sciences, and continued to write poetry—similar to his friend and colleague at the Salk, the biophysicist Leó Szilárd, who was a successful author of science-fiction stories.

It is hardly surprising, then, that the visiting scholars who came to the Salk Institute in these early years were not narrowly focused on the biological sciences. For example, Michael Crichton came to the institute as a postdoctoral student in medicine for the 1969–1970 academic year. Having published his first novels, Crichton worked with Bronowski on a project concerning the relation between science and the public.[2] Between 1965 and 1969, the leading linguist Roman Jakobson visited the Salk Institute several times. Given the recent advances in molecular biology, Jakobson's goal was to reconsider the relation between linguistics and biology. Bronowski's plan to affiliate Jakobson to the Salk Institute was never realized. However, his stays at the institute in La Jolla contributed to a number of his publications as well as prepared Jakobson's legendary appearance on French TV where he discussed the connection between "Living and Speaking" with Jacques Monod, Claude Lévi-Strauss, and others in an exemplary interdisciplinary way.[3]

A short time later, the French philosopher and sociologist Edgar Morin spent a year at the Salk Institute. It was during this stay that Morin conceived the "anthropo-socio-biological parliament" of knowledge that, drawing on systems theory and the notion of complexity, he described in greater detail in his multivolume project *La méthode* in the years that followed. While at the Salk Institute, Morin recorded his partly autobiographical, partly ethnographical observations in a notebook that, in polished and revised form, was published as *Journal de Californie* in 1970. In this book, the Salk Institute actually figures as a "monastery of science" situated in a cultural landscape where life is experimented on in various ways: in student communities, via rock concerts, drugs, and movies, through new forms of spirituality (from Billy Graham to Scientology), and through conferences and colloquia exploring the relation of biology to the social sciences.[4]

At the Salk Institute

Latour arrived at the Salk Institute in October 1975, having just completed his doctorate in philosophy. When he first set eyes on the architecture, he thought it looked like something out of a science-fiction film. The building complex, which exerted an almost magical fascination on the newcomer, Latour later described as a mixture of "a Greek temple and a mausoleum" (LLF 11).

Before his first meeting with Jonas Salk, Latour was fully expecting a modern-day Pasteur, a prophet of microbiology, and at the same time a restless researcher and manager. The philosopher was very surprised to encounter "a sage" who "talked about Picasso and the wife of the minotaur whom he was now harboring in his labyrinth" (LLF 11). Actually, this was not so surprising considering the sage was married to the ex-partner of the painter: French-born Françoise Gilot.

The physiologist Roger Guillemin was also from France; he had been working at the Salk Institute since 1970. Coincidentally or not, like Latour Guillemin was from Burgundy. In fact Latour even knew him, because as a child Guillemin had been in the choir of a local church that had been directed by a "dear uncle" of Latour's.[5]

Guillemin was born in Dijon into a family that for a time—how could it be otherwise—was engaged in the wine business. He began to study medicine in his hometown. After hearing a fascinating lecture in Paris by Hans Selye, the famous endocrinologist and "father of stress research," in the late 1940s Guillemin was successful in his efforts to move to Canada and work with Selye at the University of Montreal. There his interest soon focused on a certain group of neurophysiologically relevant chemicals, the so-called stress hormones. These hormones are produced in certain regions of the brain and are of decisive importance for our entire body's capacity to adapt physically and mentally. The principal issue in this field, which was soon to be known as "neuroendocrinology," was that of the messenger substances that trigger the production and release of stress hormones in brain regions like the pituitary gland, for example.

Guillemin devoted the major part of his research to this question—from 1953 to 1970 at Baylor College of Medicine in Houston, Texas, and then at the Salk Institute, where he was active until very recently. In 1977, the Nobel Prize in Physiology or Medicine was divided, and one half was awarded jointly to Roger Guillemin and Andrew V. Schally, Guillemin's former colleague and later competitor, "for their discoveries concerning the peptide hormone production of the brain," as the Stockholm committee put it. This referred in particular to the purification, isolation, and identification of the structure of thyrotropin-releasing hormone (TRH), a cerebral messenger produced in the hypothalamus that, put simply, stimulates the release of thyroid-stimulating hormone in the pituitary gland. The other half of the Nobel Prize went to the American physicist Rosalyn Yalow "for the development of radioimmunoassays of peptide hormones," a groundbreaking technique that was crucial for Guillemin and Schally's work.

Guillemin has said that from the very beginning he was attracted to the entire setting and environment of the Salk Institute. When he visited the institute for the first time in the late 1960s, the sight of its "monkish lines" gave him a "shock" and triggered a "spiritual experience," which he never imagined having in that particular place and has not had since, except for when he saw Cologne Cathedral at the end of World War II, "all black and how gothic and intact, in the middle of such devastation."[6] In 1970 Guillemin accepted Salk's offer to join the institute. He was given nine hundred square meters of free space, in which he could set up "a highly efficient, multipurpose laboratory" according to his own ideas.

Guillemin decided that the different areas of his laboratory would be kept strictly separate: one half of the space was devoted to physiology and the other half to chemistry; an island in the middle contained a conference room and ten small offices for his staff. The interior fittings resulted in a laboratory that was literally transparent: "All central walls were of glass so that one could see through the whole space from any one location: nobody could, or should think they were working alone in that laboratory. There were ceiling to floor length curtains, though, in the staff offices that could be closed should one wish. The curtains were rarely drawn."[7] It was here that, from the winter of 1975 to the summer of 1977, Latour gathered the raw material for a publication in the as yet young discipline of the sociology of science that would cause a sensation as the first contribution to the anthropology of daily life in a laboratory.

Laboratory Reports

Reports about visits to laboratories have probably been written ever since modern laboratories came into existence. In the 1860s, for instance, the French chemist Adolphe Wurtz traveled around the German-speaking countries on a tour of inspection of all existing laboratories for physiology and chemistry in order to report back to the French government on architecture, technical equipment, and organization. In the years 1910 to 1930 the American nutritionist Francis G. Benedict paid repeated visits to laboratories all over Europe. The extensive documentation of his findings Benedict used to optimize his own laboratory back home in Boston.

In other words, by the mid-1970s there existed a long tradition of scientifically motivated visits to laboratories and a corresponding genre of published descriptions (often, as with Latour's, featuring floor plans and photographs). However, sociologists or ethnologists making detailed observations of research practices were a complete novelty. In 1936 the Norwegian philosopher and later founder of deep ecology, Arne Naess (PN 256), suggested in an almost forgotten book that the actions of scientists, both verbal and nonverbal, should be described using the methods of behavioral science; that is, they should be observed as though by a "researcher from a different galaxy." However, only the work of the sociologist Stewart E. Perry, who in the 1960s observed the scientific practices of medical practitioners in a psychiatric clinic, and the studies on *From Anxiety to Method in the Behavioral Sciences* published by the French-American ethnologist and psychoanalyst Georges Devereux in 1967 actually went in this direction.[8]

In the French context additional arguments for observing the everyday life of scientific laboratories were provided by a somewhat surprising convergence of sociology and theology. In the early 1970s, the Jesuit historian and philosopher Michel de Certeau investigated in a series of essays, "what contemporary societies make out of religion."[9] Science was one of the examples de Certeau referred to in this context. In the heyday of structuralism, he questioned the importance of the hard sciences for understanding the totality of human reality. In the wake of Michel Foucault, de Certeau referred to language and history as the "unspoken" (*non-dit*) of the sciences and highlighted the fact that the functioning of science is always tied to "the facticity of its dependencies and the contingency of practice."[10] More generally, he stated: "There is an essential relation between the universality that any authentic science strives for and the specificity of its socio-historical situatedness."[11]

Early on, one of de Certeau's students, the theologian and former microphysicist Georges Thill, developed these observations into what he called a "praxeology of science."[12] A key element of this praxeology was a "laboratory diary" that Thill kept while working at various research institutes for particle physics. As de Certeau's biographer François Dosse has suggested,[13] this diary and its extended theological, sociological, cultural, and existential analyses concerning the "functioning of scientific production"[14] can be seen as one of the models for Latourian laboratory studies.

Despite that Latour rarely refers to de Certeau (see, however, LL1 107), his interest in observing laboratory science "in action" inscribes itself into this contemporary convergence between theology and the sociology of science. As we have already seen, Latour's parallel exegesis of *Clio* and the New Testament gospels played a decisive role in the fact that, from the start, his sociological attention to science focused on the practices of inscription and translation.

In the Anglo-American context, the sociology of science was on quite a different track in the early 1970s. Under the aegis of the sociologist Robert K. Merton the major focus of investigation was how and why it had come about that in modern societies scientific knowledge is seen as "the last resort for solving virtually every problem."[15] The principal themes were the normative structure of scientific work, the processes of assessment and recognition of scientific achievements, and the social contexts of the science system.[16] Like other pioneers of social science laboratory studies—Karin Knorr-Cetina, Trevor Pinch, or Sharon Traweek—Latour pursued his work at the Salk Institute largely independently of this kind of sociology.

Latour's approach to his laboratory study consisted essentially in being a philosopher who is interested in science. Paradoxically, it is the fact that his academic background is *not* in sociology or anthropology that enabled him to become an anthropologist of science. This is mentioned in a rather tongue-in-cheek way in the afterword to the second edition of *Laboratory Life* in 1986:

> Professor Latour's knowledge of science was non-existent; his mastery of English was very poor; and he was completely unaware of the existence of the social studies of science. Apart from (or perhaps even because of) this last feature, he was thus in the classic position of the ethnographer sent to a completely foreign environment. (LL2 273)

Guillemin's History

Roger Guillemin was not only a highly successful laboratory scientist; he also pursued historical and epistemological interests. Not least because of the ongoing rivalry with his former colleague Andrew Schally, in a longer review article published 1971, for example, Guillemin included a "precise historical account" of the single steps and results that had led to the isolation and identification of TRH in his laboratory. In 1977 he produced a similarly "historical" report concerning the isolation of another substance, gonadotropin-releasing hormone (GnRH).[17]

Guillemin's interest in history, however, was not confined to more or less polemical exchanges with rival scientists. Since the late 1950s, when he was still at Baylor College in Houston, he had published on the history of physiology together with his former boss Hebbel E. Hoff; in Guillemin's time the history of science was still an integral part of teaching future doctors and physiologists.[18]

The quality of Hoff's contributions to the history of science is exceptional. Even Canguilhem, who was known for being exacting, described an essay that Hoff had written in the 1950s on the early history of research on the reflex as "a model for the genre."[19] In 1963, together with Leslie Geddes and the young Guillemin, Hoff wrote an article on the history of blood transfusions; 1967 saw the publication of the famous laboratory notebook of Claude Bernard, the *Cahier rouge*, translated by Hoff, Guillemin, and his wife Lucienne; and in 1974 Hoff and the Guillemins produced an English translation of Bernard's *Leçons sur les phénomènes de la vie communs aux animaux et aux végétaux* (1878–1879), in which there was new interest thanks to the successes of molecular biology.[20]

And that was not all: in 1957 Guillemin published together with Hoff and Geddes an important contribution to the history of the graphic method. This could, of course, be construed as further evidence for the interplay of "primary Burgundism" and *passion de la trace*. In this essay, Guillemin and his co-authors demonstrate that long before Carl Ludwig introduced the apparatus for obtaining graphical records of organic processes into the physiological laboratory in 1847, self-registering devices similar to the kymograph were being used by meteorologists—a finding they backed up with photographs of the original device, which still exists in Paris.[21]

When Latour and Guillemin met in the early 1970s, Guillemin, who was clearly taken with the openness of the new research institute in La Jolla, invited Latour to conduct "an epistemological study" of his new laboratory, providing he secure his own source of funding (LL2 274). However,

Guillemin's interest in the history of physiology and the graphic method was only one factor that cushioned Latour's switch from the comfortably familiar contexts of philosophy and development sociology to the unaccustomed world of laboratory science. The other was the campus life of the nearby University of California at San Diego.

High-Tech, the Beach, and the Poststructuralists

It is 1975. Foucault has just published *Discipline and Punish* and goes on his first trip to California; Deleuze and Guattari are writing their eulogy to subterranean networks, "Rhizome"; and Roland Barthes brings out a book with the noteworthy title *Roland Barthes par Roland Barthes.* In other words, in France the orthodox forms of Marxism, structuralism, and psychoanalysis are in the process of exiting from the center stage of intellectual life. They are replaced by neo- and poststructuralism, which increasingly dominate the scene.

In the importation of this theoretical shift to the United States the University of California at San Diego plays a seminal role. In the mid-1970s this university is a stronghold of the academic avant-garde, of political activism, and permissive attitudes; influenced on the one side by unorthodox Marxists (Herbert Marcuse, Fredric Jameson) and on the other by exponents of "French Theory," such as Jean Baudrillard and Jean-François Lyotard, as well as Michel de Certeau, who visit on a regular basis. Beach life and surfing culture form attractive counterpoles to this mass presence of academics and intellectuals.[22]

Parallel to his daily work in Guillemin's lab Latour participates actively in the life of this "mythical campus." In the 1975–1976 academic year, for example, he attends Lyotard's lectures on Nietzsche, decadence, and the sophists. That the philosopher would shortly later be engaged on a rather unusual commission certainly did not escape Latour. In 1977, Lyotard was asked by the Conseil des Universités of Quebec's government to review the problem of scientific knowledge in the most advanced industrialized nations. The central issue that the report investigates is the effects of "information and telematic technology" on the status of knowledge in contemporary society. Lyotard's study was published in 1979—at the same time as Latour and Woolgar's *Laboratory Life*. An English translation appeared in 1984, under the title *The Postmodern Condition: A Report on Knowledge*.

One of the assertions well worth thinking about in this famous and often-cited work—which today seems to be seldom read in its entirety—is "data banks are the encyclopedia of tomorrow"; another is "the games of scientific language become the games of the rich, in which whoever is wealthiest has the best chance of being right."[23] And when Lyotard seeks to characterize "the pragmatics of scientific knowledge" he references an essay of Latour's that had just been published—the first fruit of his sojourn at the Salk Institute. The essay is about the "rhetoric of science" and is cited by Lyotard to characterize the language games of science from the point of view of linguistics. Latour's article analyzes these language games by means of semiological methods, taking a neuroendocrinology article by Guillemin and collaborators as an example.

The unequal authors—the poststructuralist Lyotard at the peak of his fame and the young post-doc philosopher in search of a research theme—meet on the common ground of their preference for questions relating to signs, language, and discourse.[24] They also agree that scientific knowledge is an "agonistic" practice; that it is a field where to speak is to fight—a position that Lyotard derives especially from Nietzsche.[25] As we shall see, the essay by Latour brings both aspects together, speaking and agonism, with regard to a concrete example of Guillemin's work.

Science as an Agonistic Field

Assistance was forthcoming from the Italian semiotician Paolo Fabbri. At that time Fabbri was associated with the circle around Umberto Eco, but his main influence was the linguist and semiotician Algirdas Julien Greimas, who taught in Paris. Latour later frequently cites Greimas when defining "actors" and/or "actants" as pivotal notions of his approach to science and technology. In the mid-1970s Fabbri was often a visiting scholar at UCSD, where he met Lyotard and Baudrillard as well as Latour and now and again attended to the social gatherings at the Salk Institute. Latour confided to Fabbri that, so far, the predominant impression of his time in Guillemin's lab was that "they fight all the time."[26] He fancied himself surrounded not by scientists but by young entrepreneurs or warriors, by "wild capitalists." At any rate, day in, day out, talk revolved around strategy, occupation of positions, infiltration of ideas, destruction of reputations, defeating opponents, and even of guerrilla warfare (LLF 12).

To portray this agonistic dimension of scientific work in the laboratory was seen as a main object of the essay by Latour and Fabbri. The focus is a four-page paper that Guillemin and co-authors published in 1962 in the venerable *Comptes Rendus* of the French Academy of Sciences. Instead of portraying science as an activity oriented on nature, in their essay Latour and Fabbri reverse this perspective. Using "overt military terminology" they depict scientific actions as a series of operations that targets a field that is identified as the current state of the literature: "In this schema nature provides the ammunition which makes it impossible to parry the blows [aimed at the existing literature]" (RdS 94). Latour and Fabbri point out that to see things from this perspective is new. Here, "persons" appear on the scene who would normally be considered entirely out of place on the science stage: "the strategy, the right, the battle, the will, and above all the rhetoric" (RdS 94).

In this connection, Latour and Fabbri allude in general ways to Lyotard and, by the very nature of the subject, also to the ideas of the literary scholar Mikhail Bakhtin, who was interested from a very early point in the rhetorical "polyphony" of literary texts. In his later work, Fabbri pursued this direction further.[27] Additionally, Latour and Fabbri follow the characterization of the field of science that the Parisian sociologist Pierre Bourdieu had recently put forward. Indeed, Latour and Fabbri's article is published in the new journal founded by Bourdieu—*Actes de la Recherche en Sciences Sociales.* In 1976 Bourdieu had written a keynote article in the same journal in which he argued that "the field of science" should not be understood as "a seemingly pure and disinterested universe."[28] In this context, he mainly referred to the Anglo-American tradition of sociology of science (Merton, Ben-David, Barnes, Bloor) while at the same time attempting to remain true to French epistemology (Bachelard, Canguilhem).

Likewise inspired by Nietzsche, in the 1940s Canguilhem, one of Bourdieu's teachers, had portrayed science as an eminently polemic activity.

According to Canguilhem, science imposes a requirement on a given, "a given whose variety, disparity, with regard to the requirement, present themselves as a hostile, even more than an unknown, indeterminant."[29] Canguilhem's description actually concerns the relationship of science to its objects of study but also applied it to science as an activity and an operation.

In a constructive as well as critical dialogue with contemporary established sociology of science, Bourdieu argued the case for treating science as one social field among many. In his view science is a field replete with internal relations of force and tendencies toward power concentration and marked by battles, strategies, and the interest in profiting academically. Just as in other sections of society, in science the goal is to accumulate "symbolic capital" and succeed in the competition to monopolize authority. Latour and Fabbri concur with this view when they state that their intention is to rediscover the combative society lying beneath its apparently peaceful rationality (RdS 95).

The advantage is obvious. Latour can continue to pursue the problem of belief, of ideology, in the science sphere as well in the form of interrogating the credit and the credibility of scientists. This is the one line that he delineates in the article written with Fabbri. The other, more innovative line is applying the tools of linguistic and semiological analysis to a text from one of the hard sciences.

The Rhetoric of Science

The starting point of the analysis is a critique of the quantitative approach to scientific literature that enjoyed a certain prominence in the sociology of science at that time. From the late 1960s onward, science researchers like Derek de Solla Price and Henry Menard had studied the development of disciplines using bibliometric data of journal titles and published articles. Latour and Fabbri set themselves apart from this quantitative approach. The philosopher and the semiotician are concerned with analyzing the specific rhetoric of scientific texts, including the "style" peculiar to them (RdS 82).

First, this opens up new territory for a method that was developed for and applied to works of literature. With the humanities and social sciences in mind Fabbri's mentor, Algirdas Julien Greimas, had taken the first steps in this direction in his book *Sémiotique et sciences sociales.*[30] While Greimas had concentrated on the social sciences, Latour and Fabbri demonstrate that scientific texts are also amenable to semiological analysis. Thus they are suggesting that scientific literature also provides a form of narration, albeit with its own, quite specific style. In their opinion scientific literature is less like a novel and more like a telephone book (RdS 88).

Second, through this approach the sociological problematic sketched above can be perceived in the details of a scientific publication. The conflict-laden society, for example, becomes tangible through the use of social markers in the text. Here Latour and Fabbri include the naming of persons, institutions, and places as well as the interspersed references to the context for producing the text, for example, the particular lab, the methods, and the instruments.

Similarly important is the issue of "modalizing," that is, the use of indirect speech, and graduated judgments and assumptions with varying degrees of probability. The specific character of a text, according to Latour and Fabbri, can be described quite precisely with reference to the number of instances and the usages of modalizing. To make this somewhat clearer using an example that the authors Latour and Fabbri do *not* use: a literary text like Thomas Bernhard's *The Loser*, a continuous first-person interior monologue that has many instances of double modalizing ("said he, I thought"), from this perspective appears to be the exact opposite of a scientific text.[31] A scientific text may contain citations, but it does not leave these as approximations and in fact pursues at the same time the opposing object of arriving at statements that are modalized as little as possible. Ideal would be "A = B," without quotation marks, without a footnote.

Latour and Fabbri not only count the instances of modalizing in Guillemin's text; they also make a distinction between the number of active and passive formulations. In addition, they observe that there are criteria for demodalizing. These are, on the one side, the standards for methods and procedures that have been specified elsewhere, partly by the same authors, Guillemin et al., and are repeatedly cited in the text; on the other side, there is the compliance or noncompliance with specific external standards that serve to establish the credibility of the reported facts and thus also of the scientists involved. These "credibilities" are established by reproducing numerical values and figures or diagrams that are generated by the laboratory's instruments.

In this connection, Latour and Fabbri speak of the "symbolic writing of the instruments" (RdS 84), which the authors of scientific texts literally place between the lines. As a variation on a formulation of Nietzsche's, Latour and Fabbri summarize their findings: "The right [of the scientist] is the way in which the will to know is translated into experimental asceticism" (RdS 91). To put it differently, the goal of the laboratory scientist is to let his or her results obtained with the aid of instruments speak for themselves.

This was a conclusion that had very little to do with the epistemological study Guillemin had had in mind when he extended the invitation to Latour. Latour made no secret of the fact that the essay written with Fabbri was a provisional work. In fact, though, Latour had discussed his drafts for the text with his "colleagues" from the lab and also with Guillemin himself before he published the article. But apparently Guillemin was not amused. In a postscript to the article the *Actes* printed some of Guillemin's comments that Latour passed on—and they are quite strong stuff: "text fetishism," "blind to the information received and the concrete features of experimenting," "a naïvely cynical and agonistic view of scientific strategies," "underestimation of the epistemological obstacles" (RdS 95).

Guillemin had expected something quite different from his philosopher-in-residence. Visibly dissatisfied, he said: "Everything is attributed to personal motives, as though scientists are shifty and cunning. But we do research, we're not playing around" (RdS 95).

This is the first time but certainly not the last that Latour will hear objections like these, sometimes from scientists, sometimes from sociologists and philosophers. Yet this did not result in an overt breach between the two men, only a distancing. Later, Guillemin only speaks about Latour in a guarded fashion, and Latour speaks of the "disinterest" (LLF Acknowledgements) shown by the famous neuroendocrinologist from the Salk Institute who made possible the field research that was presented two years later in *Laboratory Life*.

THREE

Machines of Tradition

Latour's stay at the Salk Institute ended in August 1977. He then moved to Paris, to the Conservatoire National des Arts et Métiers, to take up the position of assistant to Jean-Jacques Salomon, professor of "Technology and Society." The Conservatoire is a venerable state institution of higher education dedicated to the advancement of science and industry, and Latour was responsible for coordinating all activities in the area of science and technology research.

During this phase, Latour's activities initially focused on the Fondation Maison des Sciences de l'Homme (FSMH). Founded in the early 1960s, the FMSH is a humanities and social sciences research institute that is interdisciplinary and has an international orientation. Thus it was not surprising to find a project named PAREX at its premises on Boulevard Raspail in the early 1970s. This project had been initiated by the psychologist Gérard Lemaine in PARis and the science historian Roy McLeod of SussEX

University. The PAREX project had been set up to promote collaboration in the emerging field of sociology of science on both sides of the Channel. In addition to collaborations between France and Great Britain, cooperation was fostered with scholars from other European countries, such as Germany and Austria. Besides Lemaine and McLeod, in 1975 active organizers of the group included Michael Mulkay from York University and Peter Weingart from the University of Bielefeld. PAREX is considered by many to be the decisive nucleus of innovative social studies of science in the European context.[1]

One of the younger participants at the meetings organized by PAREX was John Law, and another was Steve Woolgar, a student of Mulkay's who was examining the development of new fields of research in contemporary astronomy. The concrete example that Woolgar was looking at concerned research on pulsars: relatively diffuse celestial objects thought to be neutron stars that emit regular pulses of radio waves and other electromagnetic radiation.[2] Mulkay had published a seminal article in 1973 with David Edge on the social, cognitive, and technical aspects of the emergence of radio astronomy, which Latour and Fabbri had quoted with approval, not least because in their article Mulkay and Edge devote particular attention to the rise of a new kind of scientific "discourse."[3] At York another student of Mulkay's, G. Nigel Gilbert, was investigating the discourses of the field of meteor radar, which, at the time, was equally new.

When Latour and Woolgar met in Paris they already knew each other. They had become acquainted at the first annual meeting of the Society for Social Studies of Science (4S), founded by Robert K. Merton, in November 1976. At Cornell University in Ithaca, New York, where the conference took place, the two young scholars had given presentations on the problems involved in the sociological analysis of scientific discourse. Latour spoke about citation counting as an important element in studying the "system of actions" of scientific papers, and Woolgar talked about the problems and possibilities of sociological analyses of scientific accounts. Unlike Latour, at that point in time Woolgar already had several publications to his credit: a review article co-authored with G. Nigel Gilbert on quantitative social studies of science; a study on problem areas and research networks in science with co-authors Mulkay and Gilbert; and a contribution to the 1976

PAREX anthology, which focused on identifying and defining topical themes in science.[4]

As these publications demonstrate, Woolgar already possessed broad and detailed knowledge of the field of social studies of science, including its methods and theories. In connection with the quantitative analysis of scientific discourses, the young sociologist not only spoke of "networks" (a term used for the first time in this context by de Solla Price in 1965)[5] but together with Gilbert also introduced a dynamic, quasi-historical component into the investigation of such networks, for example, by distinguishing between "citations" and "references" and dealing with the spinoff question as to the age distribution of references. In view of the fact that this was a relatively new research field Woolgar could also claim the need for "speculative notions" in order to narrow down and frame the potential phenomena and issues for debate.[6]

Woolgar was skeptical whether the sociology of science as it existed was capable of doing justice to the complex subject of science, and his later works show that this skepticism remained. One of the most acute issues for him was whether sociology could deliver descriptions of science that would be acceptable to scientists. Reflections of this kind Woolgar considered to be productive, integral parts of sociological research. This would also become clear in his collaboration with Latour.

However, already in the article written with Gilbert, Woolgar tentatively leaves the territory of sociology for the terrain of information science and biology. The mathematician William Goffman's pioneering use of disease epidemiology concepts to model the spread of scientific knowledge and ideas Woolgar finds especially intriguing. In Goffman's model of "knowledge epidemics" Claude Shannon seems to meet Louis Pasteur. How long will Latour resist this theoretically appealing as well as sociologically controversial mixture?[7]

Laboratory Life

The book that was the product of this constellation caused quite a sensation among sociologists and historians of science. Together with the sociological laboratory study by Karin Knorr-Cetina, *The Manufacture of Knowledge*, published in 1981,[8] *Laboratory Life* was soon being hailed as a "pioneering contribution" to science studies. Today it is recognized as a "modern classic" and, because of its programmatic significance, often compared with Thomas Kuhn's *The Structure of Scientific Revolutions*.[9]

Yet the book by Latour and Woolgar (the only joint publication of the two authors) has to be seen as a transitional work in several respects. It marks the entry of an empirically oriented philosopher from France into the international social studies of science scene. In fact, it is Latour's first publication in English. Further, with this book Latour opens up the French context, which is mostly characterized by the philosophy and history of science, or historical epistemology, for a debate with Anglo-American science studies. This coincides more or less with the opening up of the work of Lemaine and especially of Bourdieu. However, with Latour's intervention this opening up of the French scene gains much more momentum.

Fifteen years previously, Louis Althusser had enthusiastically observed with regard to Canguilhem and Foucault that these epistemologists approached the sciences with impressive expertise and in this sense "are similar to the ethnologists who go 'into the field.'"[10] In 1979 it appears that Latour could effortlessly go beyond this. He had literally spent two years "in the field," had assumed the role of an ethnologist in a laboratory, and had observed the practices of the mysterious tribe of neuroendocrinologists while they constructed scientific facts. In contrast to Canguilhem Latour did not seek to understand the details of the theoretical and methodological problems scientists were struggling with. That would have been the prerequisite for the "epistemological study" that Guillemin had in fact wanted. Instead, Latour brought with him the work of de Solla Price, Mulkay, and Edge to Paris from his stay as a visiting scholar in the United States.

Laboratory Life is also a transitional work because it does not offer any easily comprehensible point of view but lays out instead an entire range of perspectives on the phenomenon of science. And in fact the two authors, who fuse into "we," do not confine themselves to playing the role of the ethnologist or anthropologist; they also don the costume of the historian—a familiar course of action for Latour since his work on Péguy—in order to then appear as an epistemologist and, ultimately, as a sociologist. The introduction to the book is written by a sociologist, who, because he regards Merton's approach as too theoretical and aloof, advocates concrete sociological observation of scientists with results that are comprehensible to the scientists themselves (LL1 38). Here one recognizes the hand of Woolgar. The concluding chapter, however, appears to have been written by a philosopher-scientist who is just as fascinated by cybernetics and game theory as by biophysics and molecular biology. This is more Latour speaking. In fact, the end of *Laboratory Life* anticipates Latour's exploration of Michel Serres's quasi-cybernetical philosophy.[11]

At the same time, one can identify traces of earlier influences on Latour. For example, on the very last page of the book Latour includes an assumption that clearly goes back to his studies of Bultmann's work, that "the basic prototype of scientific activity is not to be found in the realm of mathematics or logic" but in exegesis—"exegesis and hermeneutics are the tools around which the idea of scientific production has historically been forged" (LL1 261). Marc Augé's ethnological analyses of the "ideo-logic" in Côte d'Ivoire are also quoted in the final pages of the book. Reference is made to Augé earlier in the epistemological chapter of *Laboratory Life* in order to describe the seemingly rigorous character of scientific arguments as a special form of "practices of interpretation." Then, on the penultimate page, Augé's work is presented as an "intellectual framework for resistance to being impressed by scientific endeavor" (LL1 260).

And there is more. As we shall see, it is above all his structural reading of Péguy's *Clio*, that is, Latour's interest in the tension-laden relationship of event, tradition, and community, that is reflected in the book written with Woolgar.

Desks Versus Machines

At the center of *Laboratory Life* is the exploration of a machinery of tradition, which via discourse and words, writing and paper, hands and inscription devices enables the establishment of a connection to events that can scarcely be comprehended by a modern, secularized notion of history.

Similar to Latour's essay on Péguy and the study with Fabbri on the rhetoric of science, the focus of interest in *Laboratory Life* is *not* the comparatively secular sphere of technology. Unlike Deleuze and Guattari, Latour and Woolgar's primary concern is not the machines themselves, the body as an "overheated factory" under the skin, or the industrial subjection of "mechanical *and* intellectual organs" to the central motive force of a factory.[12]

In the anthropologically oriented first chapter mention is made of the rotary evaporators, centrifuges, mixers, and other "machines" by the aid of which the laboratory staff of the Salk Institute cut, grind, shake, and so forth the organic material they work with. An entire page is devoted to a table listing the larger pieces of equipment in Guillemin's laboratory, which includes the name of the instrument, date of first introduction, field of origin, and usage in the program (LL1 67).

However, the main theme of this anthropological section is the writing desks upon which very different types of literature—published journal articles, computer printouts with columns of figures, diagrams, tables, manuscripts, and so on—are collected before being transformed into scientific publications. In other words, Latour and Woolgar make sure that from an anthropological point of view scientific work in the laboratory should be understood as "literary inscription" (LL1 47).

In this perspective, laboratory work is mainly and primarily continuous work on and with texts by means of which different kinds of statements are transposed into matters of fact with varying modalities:

> The work of the laboratory can be understood in terms of the continual generation of a variety of documents, which are used to effect the transformation of statement types and so enhance or detract from their fact-like status. (LL1 151)

In this connection, the authors understand "inscriptions" not only as letters and symbols. Rather, they use the term "to summarize all traces, spots, points, histograms, recorded numbers, spectra, peaks, and so on" (LL1 88). For this conception of inscription Latour and Woolgar refer to Derrida's generalized concept of writing as it had been spelled out in *Of Grammatology*, published in French in 1967.[13] At the same time, they cite François Dagognet's 1973 book, *Écriture et iconographie*, which with regard to the iconography and written language of science not only follows Derrida but also Marshall McLuhan.[14] However, the central concept of the anthropology chapter in *Laboratory Life* is not writing but "literature." The authors explain that this term

> refers both to the central importance accorded [to] a variety of documents and to the use of equipment to produce inscriptions which are taken to be about a substance, and which are themselves used in the further generation of articles and papers. (LL1 63)

In other words, Latour and Wollgar's study is about *laboratory texts*, their archiving, filing, processing, and generation, and at the same time it is about the meaning, the significance, and the validity that these texts and the writings and inscriptions used to compose them are assigned in *laboratory contexts*. To use Péguy's words one could say that *Laboratory Life* examines a specific "cult of inscription": a cult in which the inscription, if not taken for the action itself, is at least taken for the matter, the fact.

This means that, ultimately, a question of belief is again at issue: the belief not in the writing but in science. According to the authors it is the practices of interpretation and exegesis—what Augé would call "local tacit negotiations, constantly changing evaluations, and unconscious or institutionalized gestures" (LL1 152)—which inside the laboratory nurture and consolidate the "belief in the logical and straightforward character of science" (LL1 152).

That is the reason why *Laboratory Life* does not simply propose that scientific texts be subjected to classical exegesis, although the authors do cite Bultmann's *History of the Synoptic Tradition* when they attempt to explain the social context of scientific "stories" of different genres (LL1 185). Instead, the laboratory as a whole becomes an institution of tradition in which there is continual production of oral and written exegeses, rereadings, and revisions, which refer to organic and machine-based laboratory events.

Thus it would appear that here, too, Latour's vertical reading of Péguy has executed a ninety-degree turn. As we saw above, this reading revealed that *Clio* connects up the differentiating repetitions constitutive for the process of tradition with a religious event in the depths of time. Just as Latour's development sociology study in Abidjan had thrown up questions about the "circulation of information" and the "possession of documents" in actual institutions (factory, school), in *Laboratory Life* the interplay between difference and repetition also applies to processes that take place within a given space, namely, the production of scientific facts in a specific space that is part of the laboratory complex.

This is also made clear by the first figure of the book, a map of Guillemin's neuroendocrinology laboratory in La Jolla. Reading and writing of texts is done in the central Section A. Apart from the desks, there are only books, articles and essays, and reference works on terms and materials. Work with instruments, apparatus, and machines takes place in the adjacent area of Section B. Accordingly, the question the authors pose at the very beginning of their study is: "What is the relationship between Section A ('my office,' 'the office,' 'the library') and section B ('the bench')" (LL1 45)? Put slightly differently: How is the "series of transformations" constituted? By means of which events do the events that take place in Section B due to the interactions of organisms and machines finally find their way via inscription devices and other processes of reading and writing into the piles of paper in Section A—and from there, as literary end products of laboratory work, then make their way into science journals, textbooks, and anthologies? The problem of tradition in space thus becomes a question of "reference," as it is termed in Latour's later studies, for example, on the work of soil scientists in the Amazon region of Brazil around Boa Vista (CB 171–225/PH 24–79).

History and Construction

However, the way in which *Laboratory Life* answers the question as to the relationship of Section A to Section B shows (in rather Péguyesque terms) that, eventually, the boundary line between machines and desks cannot be drawn as clearly as the architectural layout of the laboratory suggests. For not only space but time also plays a role. This becomes clear in Chapter 2 of the book, when a historical case study is deployed to investigate the "construction" of a scientific fact. The fact in question was the successful isolation of thyrotropin-releasing factor (TRF) at the end of the 1960s.

With regard to this controversial development, which, as mentioned above, Guillemin had also examined historically, Latour and Woolgar demonstrate that the creation of a "new object" in the laboratory is initially tied to criteria of repetition and similarity. According to them, these criteria are decisive for whether first assumptions and initial assertions can be confirmed experimentally. In this area of relative reliability thus established, differences then appear whose observation eventually leads to the emergence of something new.

Now for Latour and Woolgar repetitions *and* differences are inextricably linked with inscriptions, and it is to the differences between inscriptions that, in this view, new scientific objects are attached. It is telling that at this point the authors invoke Bachelard's concept of "phenomenotechnique." For just as Bachelard had been adamant that instruments in a laboratory are ultimately reified theories,[15] Latour and Woolgar insist that instruments must be understood as integral components of the process of tradition. They are not interesting on account of their own materiality but solely as mediators or translators; so to speak as reifications of earlier work of interpretation (LL1 243). Consequently, the construction of a scientific fact is above all an issue of differing inscriptions: "It is not simply that differences between curves indicate the presence of a substance; rather the substance is identical with perceived differences between curves" (LL1 128). And more generally: "An object can be said to exist solely in terms of the difference between two inscriptions" (LL1 127).

With this, Latour's reading of Péguy has yet again been rotated by ninety degrees. It has been turned upside down. The differences in the written tradition no longer refer back to an event in the depths of time, nor to something that has occurred in another section of the space. The laboratory researcher acts entirely in accordance with the "correct reader" in the sense used by Péguy, who persuades this reader, through the "ex-habitual" reading of inscriptions, to start over again once more. The reader makes out of the inscriptions an event that approaches him or her in the present moment. Only the direction has changed: the distant point in time from whence the event draws nearer does not lie in the past any longer but in the future.

The repetition here has actually "interiorized" and thereby reversed itself, as one might put it with Deleuze. It is no longer conceived of as an unvarying regression to something but as a process that is open in a forward direction, a process grounded in, initiated, and continued by an event. However, this event is not something that is referred to; it is itself a reference, a sign, a difference.

At this point the historical case study in *Laboratory Life* combines with a historiographical argument. Latour and Woolgar assume that the inscription differences give rise to a new space and entail a new time: a space in which the existence of TRF has become a fact, which is now assumed as "given" in other laboratories and which they can now work with, and a time in which a world without the existence of TRF has become almost inconceivable, for scientists tend to forget the history of their innovative exegeses and project the existence of the new object into the past (LL1 107).

The history of science also participates in the construction of this new time. According to Latour and Woolgar, in historical reconstructions of episodes that lead to the production of scientific facts, science historians usually take established facts as their starting points and extrapolate their constant existence backward. They tend to describe the activities of scientists in terminology that implies the preexistence of objects or truths that only have to be "discovered" or "disclosed." Further, they often assume a simple and unproblematic relationship between signs and signified things. Consequently, for the authors it follows that "most of the time, historical reconstruction necessarily misses the process of solidification and inversion whereby a statement becomes a fact" (LL1 106).

To arrive at an adequate description of the practice of science thus requires more than a change of historiographical emphasis and focus—from the product to the process. "Rather, the formulations which characterize historical descriptions of scientific practice require exorcism before the nature of this practice can be best understood" (LL1 129). For their part Latour and Woolgar try hard to avoid all expressions that could change the character of the processes and phenomena under examination. They endeavor to find elements of a new language that does justice to the "artful creativity" of scientists. And above all they seek to garner appreciation for situations in research in which there is *no* predetermined path (LL1 128, 129, 106).

A comparable *epoché*, that is, a methodological suspension of preconceived judgments, is undertaken in the epistemological section of *Laboratory Life*.

There it is used both in the analysis of practices of construing and interpreting that the scientists working in the laboratory use to negotiate their attribution of insights and opinions to specific individuals and in the analysis of the practices they implement in their attempts to explain how scientific facts emerge and spread. In a similar way in the following sociological chapter, the focus is not the individual or collective behavior of scientist subjects; instead, "the work sequences, networks, and techniques of argument" (LL1 188) are the starting point to show that on the one hand a distinction exists between *reward* and *credit* and on the other that the allocation and assignment of *credibility* depends on having material resources such as laboratories and instruments at one's disposal, which in turn depends on legitimacy in the eyes of institutions, which disburse funds for research, and other donors. In other words, the "capital" of a scientist does not consist—as Bourdieu in agreement with Merton suggested—in symbolic recognition within the scientific community but spans—similar to what Lyotard suggested—a complex structure made up of instruments, data, theories, and scholarly publications.

Through their explicit respect for the "artful creativity" of scientists, Latour und Woolgar endow the notion of the "construction" of scientific facts with important nuances. On the one hand their emphasis on creation bluntly opposes the idea that scientific practices are about "discoveries" or "revelations"—this was one of the points that Guillemin had already contradicted in his reaction to Latour's article with Fabbri on the rhetoric of science: "We discover, we don't create" (RdS 95). On the other hand they make it clear that "construction" is not intended to be understood as methodical fabrication or programmed manufacture but rather as the emergence of something new. It is in this sense that Latour and Woolgar insist that "to say that TRF is constructed is not to deny its solidity as a fact" (LL1 127).

This hybrid status, this mode of existence of scientific objects between construction and facticity, Latour will narrow down in later works; for example, when he speaks about the antinomy that scientific facts are "experimentally made up and never escape from their manmade settings" while at the same time insisting it is essential that these facts "are not made up and that something emerges which is *not* manmade" (FR 64). Thus it is only because the objects of science are *creations*—deriving from humans but simultaneously going far beyond humankind—that they have their own history at all. Otherwise they could simply be subsumed under the history of society or history of ideas: as mere constructions or mere discoveries.

Take from Science the Idea of Science?

It is attractive to see these thoughts as a first draft of Latour's "irreductionism" and as anticipating his project of writing the history of science as a "history of things." At the same time, these reflections on the creation of scientific facts refer us back to the geography and architecture of the place where *Laboratory Life* was not actually written but to which the book has an indissoluble relationship—Louis Kahn's Salk Institute. This remarkable, symmetrical complex positioned between the land and the ocean, art and nature, knowledge and revelation has left its mark on the text by Latour and Woolgar, although the introductory chapter and particularly the concluding one seem to work against this.

In spite of breaking away from issues in epistemology, Latour and Woolgar come back to a problem that is without a doubt an epistemological one. In the concluding chapter they discuss the relationship of the sociology of science to its subject, science. "The thought of the living must take from the living the idea of the living," wrote Canguilhem around 1950 in the context of his vital rationalism.[16] In a certain way Latour and Woolgar adopt this maxim when on the last pages of the book they encourage the sociology of science to take the idea of science from science itself, that is, biology.

And that is precisely what it comes down to in this part of the book by the demonstrative taking over of terminology from cybernetics and the references to the partly evolutionary theory, partly bioinformatic concepts from Henri Atlan and François Jacob. Through the agency of Leslie Orgel and Jacques Monod these theorems circulated around the corridors of the Salk Institute while they were received enthusiastically by Michel Serres in Paris (and Stanford).

In this part of the book, Latour and Woolgar describe the laboratory as a living system in which order arises out of disorder and whose functioning can be understood in terms of a specific signal-to-noise ratio (LL1 240). Characteristics of the system include processes such as black-boxing, by means of which earlier arguments are quasi-transformed into stabilized components of the laboratory environment. Further, the living system of the laboratory is marked by chance events, the emergence of variations, and many and diverse activities of bricolage.[17] Based on the theories of Orgel and Jacob life in its entirety can be comprehended as an ordered pattern that results from disorder through selection of random mutations. Latour and Woolgar declare that, analogously, one can also understand the reality of science in this sense: as the result of a process characterized by categories such as chance, mutation, niche, and tinkering.

With this, the sociologists meet the biologists they have been observing on the level of terminology—though it seems telling that it is not the chemical-messenger world of Guillemin's neuroendocrinology that is engaged here. More important is the fact that the distance between social science and its subject, the natural sciences, which the authors had lamented with respect to a Mertonian sociology of science, now appears finally to have been annulled.

This applies both at the terminological level and to the aesthetics of the text. As Donna Haraway observes in her review of *Laboratory Life* in 1980, it is particularly through the use of diagrams and tables that "isomorphism" develops between Latour and Woolgar's text—*the text's* production of order from disorder—and the biological explanation of life.[18] In this case it is ultimately the graphic method that bridges the divide between the two cultures of the natural and social sciences. The title of the book also hints in this direction, for "life" in *Laboratory Life* can also be taken in a biological sense.

Jonas Salk was thrilled. In his introduction he does say that the authors' "tools and concepts are crude and qualitative" compared to those of biological research, but he states emphatically that "their will to understand scientific work is consistent with the scientific ethos" of laboratory research (LL1 12). With regard to Latour Salk explains that the "young French philosopher" initially began to make a sociological study of biology but along the way came to see sociology biologically: "His [Latour's] own style of thought was transformed by our concepts and ways of thinking about organisms, order, information, mutations, etc." (LL1 13). As a consequence, Latour had come to see his work as a "subset" of biological research activities, which in turn constitute a subset of life in the process of organization. Salk viewed the book as a most significant contribution to improving relations and understanding between scientists and nonscientists.

Latour, however, had the last word. At the end of the book co-authored with Woolgar he once more returns to the subject of the relationship between observer and observed, or "informants." In a personal way Latour describes how the longer he spent as an observer in Guillemin's laboratory pursuing his own work, the more confident he became about it. The more inscriptions he made and the higher the pile of paper on his desk grew, the more sure he was that there was an essential similarity between his work and that of his informants.

This similarity had nothing to do with cybernetic or biological concepts but with tangible activities. The essential similarity between the anthropologist and the laboratory scientists he was observing consisted in the fact that both were doing *craftwork*. For example, when his informants sat around a library table discussing curves, they seemed hardly different to him at all: "they pondered diagrams, putting some aside, evaluating the strength of others, seizing on weak analogical links, and so slowly constructed an *account*" (LL1 257).

What the authors of *Laboratory Life* had done was no different, and in Latour's eyes this constituted their kinship with the lab scientists. Both constructed a report, an account, a *récit*. And in this sense Latour and Woolgar did not contribute to a new form of biological sociology with their book; instead, to quote their very last sentence, they took "a first tentative step towards making clear the link between science and literature" (LL1 261). This is precisely the context in which significant portions of Latour's oeuvre will develop and operate in the years ahead.

FOUR

Pandora and the History of Modernity

"The first bachelor machine was Pandora," wrote Jean-François Lyotard in his 1975 contribution to the catalogue of Harald Szeemann's famous exhibition on the *machines célibataires*. In the text, which he later included in his book on Marcel Duchamp, Lyotard recounts his interpretation of the Greek myth: Zeus had commanded the creation of the first woman, Pandora, because he was angry that fire had been stolen from heaven; Prometheus had stolen it to give to man, who made much purposeful use of it. So Zeus ordered Hephaestus, the god of craftsmanship, fire, and the forge to make a "machine-woman" out of water and clay. Hermes, the god of travelers and orators, among other things (Lyotard refers to him as "the wily god of change and transition"), gave Pandora her name and the power of speech. However, it is not Prometheus but his brother Epimetheus who becomes the object of Zeus's revenge. In Pandora Epimetheus, "he-who-understands-afterward," was confronted with a machine that was not a tool or a weapon but a trap—in a twofold sense. On the one hand Pandora's external appearance tricks the forces of nature by locking them in an automaton, and on the other she deceives both the Titans and men: she captures them with the power of illusion and makes them believe that the machine is an organism.[1]

According to Lyotard, the Pandora myth is not only about the relationship between gods and humans, between nature and technology. At issue is also the relationship between men and women. It is through his susceptibility to the beauty of the First Woman that Epimetheus brings forth the division of humankind into male and female, which Duchamp will put on display in *The Large Glass*. At the same time, this reinforces the division between gods and humans. As Lyotard points out, Zeus's actions in the given constellation do not reveal him to be omnipotent; rather, he reacts to the Promethean machinations of humankind. Thus, between the gods and men has long existed a partition that separates them. Not until the advent of the Sophists, those Nietzscheans *avant la lettre*, will there be a philosophical answer to the existence of such partitions: "Every argument must be opposed by another rigorously parallel one, but leading to a contrary conclusion: that sophistry is first of all the art of arguing cases with duplicity, *dissoi logoi*."[2] Only in this way can the natural and the artificial, female and male, gods and mortals come together again.

Pandora Years

Pandora is the iridescent signum under which Latour's work unfolds in 1980s Paris. This is to be understood quite literally: *Pandore* was the name of an initiative he co-founded that endeavored to introduce the new approaches of science and technology studies to France. Besides Latour, another leading member of the initiative was the sociologist Michel Callon. Trained as an engineer, since 1968 Callon had worked at the Centre de Sociologie de l'Innovation of the École des Mines. Callon and the founding director of the center, Lucien Karpik, a sociologist of economics influenced by Alain Touraine, were particularly interested in the "logics of action" of different actors in commercial enterprises as well as investigating the "networks" that exist between firms and other organizations.[3] Around 1975 Callon's interest started to focus on the sociology of science. Initially his points of reference were Bachelard and Canguilhem, but he soon gravitated to the Anglo-American sociology of science à la Merton and the related contributions by Bourdieu.

Together with Callon Latour published the news bulletin of the Pandora initiative. From November 1978 *Pandore* appeared with the aim of facilitating the planning, carrying out, and evaluation of activities in the field of science and technology studies in the Paris metropolitan area. Besides announcements of forthcoming events, book reviews, commentaries, and readers' letters, this *bulletin de liaison* also carried short essays by Callon, Latour, and many other authors, including Michel Serres and Françoise Bastide. Up to 1983 twenty-five issues were published in which Latour, for example, reviewed publications by Alfred Sohn-Rethel, Ludwik Fleck, and Elizabeth L. Eisenstein.

In Rue de Chevreuse, not far from the École des Mines, *Pandore* had its own offices. The premises were also used for seminars and as a meeting place for the members of the *Pandore* registered society whose aim was "to promote investigations, studies, publications, discussions, and seminars on the connections that exist between the development of science, technology, and society."[4] A book series that Latour and Callon edited was also called Pandora. In 1982 they published *La science telle qu'elle se fait*, an anthology of texts in sociology of science from the Anglo-American context. The volume contained programmatic contributions by Michael Mulkay and David Edge, which Latour was well acquainted with from his collaboration with Woolgar, as well as more recent work by Steven Shapin, Harry Collins, and Trevor Pinch.[5]

A further volume in the series was the French translation of David Bloor's *Knowledge and Social Imagery* (1976), one of the key texts of the so-called strong program of the sociology of science.[6] Bloor introduces the principle of impartiality, according to which the sociology of scientific knowledge must examine true as well as false scientific statements; that is, it has to investigate their rationality *and* irrationality, their successful *and* their unsuccessful claims. A further essential element of Bloor's strong program is the principle of symmetry, which Latour will later take up. This principle states that the style of sociological explanations of knowledge should always be "symmetrical"; that is, the same types of explanations should be used to account for both correct and incorrect convictions. However, for Latour symmetry will come to mean something different, namely, equal consideration of people and things, "human" and "nonhuman actors" in anthropology.

Another volume that appeared in the *Collection Pandore* series in the early 1980s was Latour's first book as a single author—his study of Pasteur titled *Les microbes: Guerre et paix*, which was published together with the philosophical treatise *Irréductions* in 1984. The full title of the work is *Les microbes: Guerre et paix, suivi de Irréductions*, whereas the English edition is simply the witty *The Pasteurization of France*. Pandora is also well represented in Latour's second book, which developed out of this context and at this time: *Science in Action* not only casually mentions the fictional substance "pandorin," but the book's introduction also has the programmatic title "Opening Pandora's *Black* Box" (SIA 153–154, 1–17).

Three years previously Gilbert and Mulkay had published a sociology of science study entitled *Opening Pandora's Box*, in which they continued their analysis of radio astronomy's scientific discourses. In this connection, Pandora's box was a metaphor for the scientists' "remarkably diverse accounts" of their actions and beliefs.[7] Whereas Gilbert and Mulkay were endeavoring to highlight the importance of this diversity of discourses for the sociological study of science, Latour again means something different. When he talks about opening Pandora's black box he is using the vocabulary of cybernetics, as in *Laboratory Life*, to draw attention to the constructed nature of facts and machines as the subject of his sociological and historical study. Thus the beakers, jugs, and containers created and used in scientific and technical practice are the issue here—Pandora as an automaton—and not the duplicitous arguments that escaped because a container was opened.

As we shall see, the reference to Pandora has even more facets. When Latour describes this mythological figure as a hybrid creature—"half virgin, half Cyborg"[8]—he is prefiguring the necessity of adopting a double, stereoscopic view, which simultaneously considers knowledge *and* belief, nature *and* society, modernism *and* post- or premodernism as it will be fully elaborated in his concept of symmetrical anthropology. Thus Pandora stands for the need to formulate two views on every subject and to discourse on every topic twice. Insofar as Latour meets these requirements his philosophy can be understood, with that of Lyotard (and Nietzsche), as profoundly *sophistic*.[9]

The Pasteur Project

Notwithstanding the many and diverse activities set in motion by the Pandora initiative and the drive and reach of *Laboratory Life*, it was a fact that in early 1980s Paris science and technology studies was still mainly influenced by the dominant tradition of historical epistemology. Althusser and Foucault continued to dominate philosophy and sociology as well as the analysis of science and history. Although Canguilhem had retired, his institutional influence remained considerable. He also published new contributions to the history and philosophy of the life sciences.[10] "*Le Cang*," as his students respectfully called him, thus remained for further generations of students the dominant figure that he had been since the mid-1950s when he succeeded Bachelard.

Jean-Jacques Salomon, with whom Latour worked at the Conservatoire National des Arts et Métiers, had also been a student of Canguilhem's. First working as a science journalist, Salomon then focused on science and education policy (also for the OECD). In the early 1970s he published various works on the relations between science, technology, and society. In 1978 he was appointed to the Chair of Technology and Society at the Conservatoire, where he set up a course of studies in this field. This was one part of the institutional background to Latour's Pandora initiative and the publications associated with it.[11]

Another student of Canguilhem's was the science historian Claire Salomon-Bayet, who at that time had been working at the Centre National de la Recherche Scientifique (CNRS) for some years as an independent researcher. In 1979 Salomon-Bayet published a major study on the history of biological experiments in the seventeenth and eighteenth centuries, which Latour and Woolgar cited with approval in *Laboratory Life* (LL1 42).[12] Next, she had embarked on a project centered on the bacteriologist Louis Pasteur and the revolution that bears his name.

Apart from his work at the Conservatoire and in the Pandora initiative, from 1978 to 1981 Latour's activities were mainly devoted to this project. He wrote contributions for the collective around Salomon-Bayet and worked on his own book on Pasteur. This is the other part of the institutional background to the Pandora initiative.

Salomon-Bayet had chosen for her project one of the most important scientists of nineteenth-century France. Pasteur, the founder of stereochemistry, developer of many vaccines (including anthrax and rabies), and inventor of the process to superheat food to kill certain pathogens and preserve it (pasteurization), is immortalized in untold French paintings and many films. Almost all cities and large towns in France have roads named after him and/or a Pasteur Square, and since the foundation of the first Pasteur Institute in 1887 a worldwide network of institutes has grown up with branches in Brussels, St. Petersburg, Tehran, and many other places, including in Abidjan.

However, this is not what really matters about Latour's encounter with Pasteur. It is also not important that Pasteur engaged with the processes of winemaking, or that he can be seen as a "forerunner" of the microbiologist Jonas Salk, or even that he is, as Latour notes, an emblematic figure for the "scientific religion" of the nineteenth century (MI 12). What is crucial for Latour is the fact that Pasteur played an essential part in constituting what we understand today as "modern society": an enormously complex structure, which is shaped by industry, technology, and science, as well as "biopolitics," that consists in the main in partially private and partially public hygiene. That this society is not modern in the sense usually assumed is one of the points Latour will put forward in *Les microbes.* Actually, one could say that his book develops the irony that lies in the fact that a decisive step was taken toward the modernization of society in the nineteenth century thanks to a man whose name has a comparatively nonmodern meaning. In French, *pasteur* means pastor or shepherd. Despite or perhaps because of these connotations, Pasteur managed to become an exemplary modern figure, a successful fighter of infectious diseases that were the scourge of the nineteenth century.

Salomon-Bayet was perfectly aware of the scientific, social, and cultural importance of Louis Pasteur. The exemplary modernity of the microbiologist and hygienist was precisely what interested her. At the same time, it was clear to her that she was working on the continuation of a theme that the historical epistemologists had already treated at length. In 1967 Dagognet had examined Pasteur's theoretical orientation on chemistry, and in the mid-1970s Canguilhem had described the effect of bacteriology on the "medical theories" of the nineteenth century.[13] According to Canguilhem, it was not Claude Bernard but Pasteur who had played a decisive role in turning medicine into a science. Besides *The Birth of the Clinic*, described by Foucault, it was the search for microbial pathogens that was the decisive factor in establishing the new, scientific model of medicine. Canguilhem describes this model as "one of knowledge without system, based on the collection of facts and, if possible, the elaboration of laws confirmed by experiment."[14]

Salomon-Bayet takes this result as her starting point but turns it in another direction. Her project, sporting the rather ungainly title "Contribution to the Social, Economic, and Political History of the Development of Scientific Research in France Since the Mid-Nineteenth Century and Particularly in the Twentieth Century," sought to combine a history of concepts à la Canguilhem with the social and cultural history of scientific practices and institutions. Pasteur and his scientific work were meant to be understood as a "global phenomenon" that the myth of "pure" science is inadequate to capture and that cannot be reduced to familiar social processes. More specifically, it was about the issue of hygiene in the French context pre- and post-Pasteur, the scientific and public debates about Pasteur's research work, its influence on the legal status of doctors and pharmacists, and obviously also the role of the laboratory as a paradigmatically modern institution.[15]

The inclusion of social and cultural history in the project's scope was, first, a reaction to the discussion in France that social contexts needed to be taken more into account, a discussion that had mainly been initiated by science historians from the United States. A particularly striking example was a case study on the debate between Pasteur and Pouchet by John Farley and Gerald Geison, who suggested that Pasteur ultimately won the dispute not because of better research findings or arguments but because of external political and cultural power relations. Latour had this essay translated into French, included it in the 1982 anthology *La science telle qu'elle se fait* that he co-edited with Callon, and at a later point in time reexamined the Pasteur–Pouchet debate himself in depth.[16]

Second, Salomon-Bayet's approach is in the tradition of Foucault's historical work, which analyzed the role of institutions, such as the clinic or the prison, in modern society. For a moment it looks as though Latour, who was so well acquainted with the contemporary conditions of the laboratory, might be interested in delving into its history in a manner oriented on Foucault and within the framework of Salomon-Bayet's project. This impression is reinforced by the fact that the idea of the "actor-network," which was beginning to take shape in Latour and Callon's thinking at this time, is a lot closer to Foucault's concept of the *dispositif*[17] than one would think at first sight. In an interview in 1977 Foucault did actually define the *dispositif* as a "network," a "system of relations" that is established among diverse components within a heterogeneous ensemble: "architectural forms, regulatory decisions, laws, administrative measures, scientific statements, philosophical, moral and philanthropic propositions—in short, the said as much as the unsaid."[18]

But as we shall see Latour was not interested in any Foucauldian archaeology of the biological gaze—far from it. Similarly, his attention was not aroused by a history of the human body and the hygienic *Care of the Self.*[19] In fact, it appeared that Latour was not even interested in pursuing the anthropology of the laboratory that he had begun with Woolgar in a historical direction. At any rate Latour's *Pasteurization* does not reproduce any floor plan of Pasteur's laboratory, it does not list any inscription devices, and even desks or exemplary publications by Pasteur himself are not a focus.

"*Give me a laboratory . . .*"

With its unorthodox mixture of sociology, history, and philosophy, its wealth of theoretical references, and its often very personal, almost creedal tone, *The Pasteurization of France* is certainly an unusual contribution to the history of science. Following on from Lyotard, from a formal point of view one can compare this book to Duchamp's *The Large Glass*. The artwork consists of two parts that are separate—below the Bachelors, above the

Bride—and Latour's book also has two sections that offer two different views. The first is empirical, and the second theoretical. Latour begins by offering a historical account, which is followed by a succession of philosophical propositions, commentaries, and reflections: first, we find ourselves, so to speak, in the "cemetery of uniforms and liveries" (history), and then we enter the "Milky Way," the territory of cloudy "flesh" (philosophy). Also, that Duchamp related the two parts of his *Glass* to each other in the sense of a projection is found in Latour's book on Pasteur. According to the author, the first part is a "map" drawn according to the principles given in the second part (MI 5).

The framework that holds the book's two parts together is, in the first place, a glossary at the end. It gives the key terms that appear in *both* parts of the book, for example, "actor," "network," "inscription," "laboratory," as well as "translation" and "force" (respectively, "weakness"). Then it turns out that the two parts of the *Pasteurization* book have the same philosophical point of reference, namely, Spinoza. His philosophy is the glass, so to speak, on which Latour draws his map. In the first part he refers to Spinoza's *Tractatus Theologico-Politicus*, a work known to him since his studies with Malet as one of the first works of modern biblical exegesis. And indeed it is the declared aim of *The Pasteurization of France* to found a new "exegesis of scientific texts" (PF 7).

Conversely, the second part of the book, with its scrupulously numbered paragraphs reminiscent of Wittgenstein's *Tractatus Logico-Philosophicus*, refers to Spinoza's *Ethics*. Here it is the steady sequence of propositions and demonstrations as well as the sudden insertion of scholia and other comments to which Latour is committed. In Deleuze's view the central question of the *Ethics* is about the body and its ability to "affect" and to be "affected."[20] Latour addresses this idea but poses a different fundamental question: How is a society put together? How is it "composed"? And what does it mean to act within a society, to become an actor (PF 9–12)?

It does not come as a surprise that in the answers to this question, which the two parts of the book provide, no individuals or collectives, no Pasteur or environment, no science or society are found whatsoever. Like in a Spinozan ethology it is all about "individuating affective states of an anonymous force."[21] In other words, what Latour is aiming at is a kind of societal "plane of immanence"; however, the way there traverses historical terrain.

In fact, *The Pasteurization of France* is first and foremost devoted to reconstructing the social and cultural revolution that Pasteur's bacteriology set in motion. To apprehend the dimensions and quality of this revolution, Latour does not examine Pasteur's writings but instead three science periodicals associated with him: *Revue Scientifique*, *Annales de l'Institut Pasteur*, and *Concours Médical*. Latour is not looking for yet another discussion of Pasteur's achievements or individual accomplishments (as Dagognet and Canguilhem had done) but rather an overview of what hygienists, medical researchers, and disciples of Pasteur had investigated and undertaken in the period 1870 to 1919.

Latour's argument follows on from the sociology chapter in *Laboratory Life*. He returns to the argument elaborated there that the allocation and distribution of credit in science is dependent on access to a complex arrangement of instruments, data, theories, and publications. Here he is not only advancing this argument against Bourdieu's position but also against historical epistemology. Taking issue with Canguilhem, Latour states that to understand the "revolution" associated with Pasteur's name it is not enough merely to refer to epistemological breaks in the development of a theory or simply to maintain that this has led to the formation of new scientific concepts (PF 31). Rather, the focus should be shifted to the social processes whereby knowledge is generated, established, and above all disseminated in a particular society.

In place of the image of a scientist who stands alone in his laboratory and revolutionizes his environment through "the power of his mind alone" (PF 14), Latour wants to turn the spotlight on the translations and the discernible transformations of the microbiological knowledge of Pasteur:

> An idea, even an idea of genius, even an idea to save millions of people, never moves of its own accord. It requires a force to fetch it, seize upon it for its own motives, move it, and often transform it. (PF 16)

The decisive authority that provides this force is clearly the laboratory. Pasteur used this type of research facility to subject the vague reports and claims of the traditional hygienists to rigorous rationalization. Hazy expressions, like "contagium," "miasma," or just simply "dirt," he transformed into references to specific pathogens on the basis of his experimental practice. However, the laboratory was also the vehicle by which Pasteur introduced his findings to society and disseminated them. The most famous example of this "Theater of the Proof" (PF 85–87) was the public experiment in 1881 in which animals, mainly sheep, were vaccinated against anthrax on a farm in Pouilly-le-Fort. The interest of the skeptical farmers and veterinary surgeons was aroused, and they were ultimately won over by the fact that the scientist had left his laboratory and performed on-the-spot vaccinations that proved very successful.

For Latour this is but the first stage in a whole chain of translations, transformations, and transportations. The second consisted in using isolated strains of anthrax to reproduce the disease in laboratory animals, not only per se but in its different forms and progression. The third stage was delivery of an effective vaccine to veterinary surgeons who were then able to fight the disease on a large scale. When all stages of this process had been completed, the laboratory was finally established as a true Archimedean point from whence French society of the late nineteenth century could be lifted off its foundations. The movements and changes that occurred on the way from this social constellation into the laboratory and then back again into society were precisely the lever that allowed Pasteur to accomplish this.

Latour stresses that this is more than just the spreading or "diffusion" of Pasteurian knowledge: rather, it is a "translation," transfer, and transport, which on both of the pathways—from society into the laboratory and then back again—changes what is being transported and what is transporting it (PF 13–16, 59–62). This is a point on which Latour is in agreement with Callon, who a few years previously had begun to focus on the "action of translating [*opération de traduction*]" as a sociological theme par excellence. Callon illustrated the prototype of this action with the following argument taken from an institutional context:

> To solve Problem A it is necessary to find a solution for Problem B because a precondition for solving Problem A is the overcoming of these or those difficulties which are associated with a solution for Problem B.[22]

Thus translations must be understood as negotiations "within a symbolic universe." According to Callon, in fact every operation "that changes a particular problematic statement into the language of a different particular statement" has to be seen as a translation.[23]

In the *Pasteurization* book Latour aspires to go beyond the sphere of mere statements. In his view it is not only the biomedical knowledge of the nineteenth century that was changed by Pasteur; the transformation that modern society also underwent was inextricably linked to this. Whether it was the packaging, transportation, or sale of perishable foodstuffs; the daily hygiene in schools, hospitals, and hostels; or combating tropical diseases, the social sphere no longer only encompassed the world of the workers and their factories but also the domains of microbes, bacteria, and viruses: "There are not only 'social' relations, relations between man and man. Society is not made up just of men, for everywhere microbes intervene and act" (PF 35).

For this reason the meaning of "society" is now no longer restricted to the global phenomenon of capitalism but embraces local mixtures of "urbanism, consumer protection, ecology (as we would say nowadays), defense of the environment, and moralization" (PF 23). Latour even goes so far as to say that Pasteur and his allies "redefined the social link" (PF 39).

Sociology and Bacteriology

Latour in his book on Pasteur confronts microbiology with the same openness as biophysics at the end of *Laboratory Life* and again uses the same historiographical arguments that he first formulated together with Woolgar. As described above, in the history chapter of their study the two authors had sought to establish a historiographical concept, the instrumentarium, that would do justice to the "artful creativity" of scientific practice. From a similar motive the study on Pasteur distances itself from mainstream historical scholarship.

Although it is obvious that in *The Pasteurization of France* Latour is dealing with historical material, he makes it clear from the beginning that he is not functioning as a historian. He says: "I cannot claim [the honor] of being a historian. This undertaking does not purport to add anything to the history of science, still less to the history of the nineteenth century" (PF 12). Latour also admits that he cannot operate like a scientist; just as Spinoza needed to become an agnostic as far as the biblical texts were concerned in order to write the exegetic sections of his *Tractatus Theologico-Politicus*, the

Pasteur scholar has to suspend all belief—in science as well as in history—when studying his sources with the aim of grasping the actual relationships between cultural and natural actors. The Pasteur scholar has to become an agnostic. This applies both to historical scholarship and to sociology. What Latour wants to do is to approach the relationship between science and society in a way that has as few presuppositions as possible—to attain a quasi-ethnological description and explanation that does not resort to using any of the terms of the tribe that is being studied (PF 8–9).[24]

Thus it seems only fitting that *The Pasteurization of France* does not begin by looking at science, history, or sociology but instead by paraphrasing a passage from Leo Tolstoy's novel *War and Peace*. For one thing, this allows Latour to return to a theme from the end of *Laboratory Life*, namely, the anti-Bachelardian proximity of science and literature. For another, he can also revisit Lyotard's portrayal of the field of science as highly polemical and agonistic, a field in which to speak is to fight. And in addition he is able to shift the problem of the actor to center stage. Like Tolstoy Latour poses the epic question as to which forces and powers enabled Pasteur to celebrate such triumphs in the field of science, as Napoleon did in his military career. How did Pasteur become such an important historical actor?

We can discern a pattern here. In his study of Péguy's literature Latour chose an extremely "nonliterary" approach. In *Laboratory Life* he was scrupulous about not letting the terminology of historiography and epistemology distort the processes observed in the practice of science. In his book on Pasteur, too, Latour begins by creating a distance and later turns this into proximity. The approach he selects in the *Pasteurization* book is extremely "nonscientific" in order to do justice to science as a phenomenon.

This procedure recalls Edmund Husserl's phenomenological method of *epoché*, an association that Latour underlines when he discusses the question of which concepts can be used to describe science:

> I must admit that there is no established stock of such concepts [that would make our explanation independent of the science under study], especially not in the so-called human sciences, particularly sociology. Invented at the same period and by the same people as scientism, sociology is powerless to understand the skills from which it has so long been separated. Of the sociology of the sciences I can therefore say, "Protect me from my friends; I shall deal with my enemies,"

> for if we set out to explain the sciences, it may well be that the social sciences will suffer first. What we have to do is not to explain bacteriology in sociological terms but to make those two logoi once more unrecognizable. (PF 9)

This delineates Latour's intention, which in the second part of the book, the "Irreductions," is elaborated and justified at length: Latour seeks to situate himself beyond any concurrence or opposition between the natural sciences, social sciences, and humanities so that he is not obliged to reduce *science* to any other quantity: not to society, not to economics, not to language games, not to the law, not to mechanisms, and not to systems (PF 203–206).

However, what Latour actually does is something different. Through studying historical texts, he tries to shift the two disciplines of sociology and bacteriology to a zone of "indiscernibility"[25] in which it becomes possible again to think about science. Latour aims at a renewal of sociology (of science) inspired by bacteriology:

> Just as the Pasteurians reshuffled the distribution of actors between nature, science, and society by the temporary formation of the microbe-whose-virulence-one-can-vary-in-the-laboratory, we [the sociologists] must, to survive, redistribute one more time what belongs to nature, the sciences, and societies. (PF 149)

As fully considered and innovative as this approach may sound, the first thing it actually does is turn Lyotard's Pandora sophistry on its head. What Latour is constructing is not two parallel discourses that arrive at opposite conclusions but two opposing *logoi* that converge in similar, if not symmetrical, conclusions. *The Pasteurization of France* centers on a sociological bacteriology that is at the same time a bacteriological sociology.

FIVE

Of Actants, Forces, and Things

Latour's conviction that Pasteur and his allies had "redefined the social link" was clearly the result of his focused analysis of late nineteenth-century biological and medical journals. Yet one can also read this statement as the consequence of a conscious decision with political undertones. Or at least this has been suggested by the British science historian Simon Schaffer in a review essay of *The Pasteurization of France*.

Schaffer finds it somewhat remarkable that Latour identifies with Pasteur to such a degree and that he never, or very seldom, quotes prominent opponents of Pasteur, like Robert Koch, for example. Further, in Schaffer's opinion, extending the concept of society to include organisms, which Latour, following Pasteur, subscribes to, seems to indicate a backward-looking, conservative element. For ultimately, extending the concept of society in this way will lead to a relapse into "hylozoism," the doctrine that all matter has life, and divert attention away from the liberating and/or oppressing potentials of human action.

Referring in this connection to Pasteur's support for Napoleon III and to a famous text by Karl Marx, Schaffer gave his review the title "The Eighteenth Brumaire of Bruno Latour." According to Marx, those who profited from the 1848 Revolution had borrowed "names, battle slogans and costumes" from the revolutionaries of 1789, and Schaffer accuses Latour of similarly borrowing a "time-honoured disguise" from the nineteenth century in order to stage a revolt against traditional sociology of science—a tradition that in Schaffer's estimation is in large part Marxist.[1]

Latour's historical and theoretical linking to Pasteur, which is apparent in *The Pasteurization of France*, can also be seen as his continuing engagement with the philosophy of Michel Serres, which began in *Laboratory Life*. In fact, he dedicates the book on Pasteur: "To Michel Serres and to all of those who are crossing his Northwest Passage." The fifth volume of Serres's *Hermès*, with the subtitle "The Northwest Passage," appeared in 1980, and shortly afterward an appreciation of the work appeared in the *Pandore* bulletin.

Latour was particularly impressed by this book. In his view, *Hermès V* provided a necessary link between the frequently cited two cultures: that is, "the firm lands of science" and "the soft resources of the humanities" (EC 90). One of Serres's following books, *Statues*, published in 1987, similarly impressed Latour. In spite of the purely philosophical character of this "Book of Foundations," he saw it as a decisive step toward a "history of things" informed by science and technology studies, which would also be a history of assemblies and, therefore, a bridge between early and late modernity. Thus Latour quotes from the work:

> In all the languages of Europe, north and south alike, the word "thing," whatever its form, has its root or origin in the word "cause," taken from the realm of law, politics, or criticism generally speaking. As if objects themselves existed only according to the debates of an assembly or after a decision issued by a jury.[2]

With regard to the alignment of science and society that is apparent here, Latour says of Serres: "*he is not modern*."[3] A few years later Latour will write, "*We* have never been modern . . ."

These references may be rather surprising considering that like Claire Salomon-Bayet, Serres also belongs to the tradition of historical epistemology that was so strong in the Parisian context. In the early 1950s Serres had studied at the École Normale Supérieure, and his thesis, on a theme from the history of mathematics, was supervised by Bachelard at the Sorbonne. Next, Serres studied Leibniz, who is often considered to be an important mediator between French and German philosophy and is also frequently cited as one of the forefathers of cybernetics.

In the 1960s Serres began an intense dialogue with Foucault, which initially revolved around *Madness and Civilization* and then, at the University of Clermont-Ferrand, around *The Order of Things*. In 1963 Canguilhem still referred to Serres as a "young epistemologist" who would facilitate a fitting characterization of the role that the history of science played in Bachelard's epistemological work.[4]

On account of his mathematics-based theory of networks in the 1970s, the same Serres was hailed as "the only philosopher in France to abide the structuralist method." For this very reason he was perceived at the same time as someone who was able to realize that the discourse of "pure" science was a myth and in his own writings systematically obliterate the differences "between the *learned text* and the *fictional text*."[5]

However, it was not Serres's structuralist network theory that Latour took up in his collaboration with Callon or in his book on Pasteur, nor was it the dialogue that had begun between Serres and the authors of the highly successful *Order out of Chaos*, the winner of the Nobel Prize in Chemistry Ilya Prigogine and the Belgian philosopher Isabelle Stengers.[6] To begin with what attracted Latour was that in spite of his connections with historical epistemology, Serres could be considered as Bachelard's "opposite number" (NB 93).

From this point of view it was to Serres's great credit that in the *Hermès* books he had rejected the belief in "epistemological breaks" in the history of science as well as in the strict separation of science and literature. Whereas Bachelard keeps his literary criticism (e.g., *The Poetics of Space*) entirely separate from his historical and epistemological work (e.g., *The Formation of the Scientific Mind*), Serres by contrast seeks to mix the two fields. For instance, he analyzes the novels of Jules Verne for their scientific and theoretical potential, interprets a painting by William Turner as illustrating Carnot's thermodynamics, or reads contemporary molecular genetics as a "translation" of the traditional motif of the tree of life. Latour summarizes this as follows:

> Instead of believing in divides, divisions, and classifications, Serres studies how *any* divide is drawn, including the one between past and present, between culture and science, between concepts and data, between subject and object, between religion and science, between order and disorder, and also of course, divides and partitions between scholarly disciplines. Instead of choosing camps and reinforcing one side of the divide, of the crisis, of the critique—all these words are one and the same—Serres sits on the fence. (EC 93)

Against the backdrop of this "Enlightenment Without the Critique" and after his studies of Pasteur, Latour is especially interested in the theory of the collective that Serres develops in *The Parasite.* On the surface this book from 1980 is a prime example of the obliteration of differences referred to above. Serres starts from the fables of Jean de La Fontaine. Obviously, he is not interested in any conventional reading of these stories about animals but instead interrogates them as to their scientific, sociological, and ecological content. At these deeper levels Serres seeks to relocate the categories of communication processes, which he had developed through his study of cybernetics, information theory, and biophysics, to a real, living environment.

In the first volume of *Hermès* Serres had declared that "following scientific tradition," *noise* is to be understood as the totality of the "phenomena of interference that become obstacles to communication."[7] In *The Parasite* he returns to this explanation but contextualizes it with regard to the small groups of animals, humans, and things that populate La Fontaine's fables.[8] The function of noise is tracked down in a variety of ecological and the attendant social constellations; for example, noises made by humans that scare away rats in the middle of a meal, but also with concrete reference to the figure of the scrounger and freeloader.

The hypothesis, which is constantly being tested here in the light of technology, economics, and society, is that in any process of communication the presence of an interfering third entity is a requisite. Thus for Serres noise is by no means a background phenomenon but a central dimension of every act of communication: without noise, no signal; without chaos, no order. It is this idea that at the end of the book carries over into a theory of the collective.

At its center is the concept of a "quasi-object," which in turn refers to a central point. Serres explains this using the model of ball games: "the game doesn't need persons," he says with sports like rugby or football in mind; it is not the players who control the game but the ball. When it circulates among the subjects in a game, "the ball isn't there for the body; the exact contrary is true: the body is the object of the ball; the subject moves around this sun."[9] In his view a similar central position is taken up by the role of the third entity—the interference in communication.

Perhaps it is not a coincidence that Serres uses a Spinozan formula to summarize the character of this relationship: "Playing is nothing else but making oneself the attribute of the ball as a substance."[10]

Latour, whose book on Pasteur as we saw also connects with Spinoza, takes up these ideas particularly in his theory of the "hybrids" that he develops in *We Have Never Been Modern*. Yet it is not only the AIDS virus and the ozone hole that appear in place of the mixtures and assemblies that an anthropology of science grounded in Serres should address. As we shall see, printed images or technical objects can also be found at this intersection. However, in *The Pasteurization of France* this focus on the Middle Kingdom is not yet prominent. By contrast, the ecological perspective that Serres develops in his study of La Fontaine with regard to humans, animals, and things is particularly conspicuous. Latour and Serres meet here in a theory of society that is inspired in the broadest sense by microbiology.

Actors and Actants

Latour changed the discipline of sociology by bringing it together with bacteriology. Something similar happened to another discipline, which Latour cites at the beginning of *The Pasteurization of France* in spite of his insistence on and claims to agnosticism: exegesis. If one follows Latour, then in this case, too, established terminology should not stand in the way of understanding and explaining the activities of Pasteur and his allies "from within" (PF 114). With regard to his own method of proceeding, in the first part of the book Latour describes the work that is necessary to achieve this as that of an analyst whose methods come in part from watching microbiologists:

> These storytellers [i.e., the Pasteurians] attribute causes, date events, endow entities with qualities, classify actors. The analyst does not need to know more than they; he has only to begin at any point, by recording what each actor says of the others. (PF 10)

Naturally, the writers alluded to here are also actors. As Latour will do later, at this point one can already quote the sociologist Erving Goffman to record this first, comparatively trivial meaning of *actor* (RS 46). The biologists and medical practitioners who fill the pages of the *Concours Médical* journal are historical actors; that is, people who act. The entire problem, but also what is interesting about this, lies in the fact that these people cannot present themselves alone in everyday life, as Goffman would say. Their social interaction is dependent on involving other actors, who simultaneously they cause to move, to shift, and to change, like in a translation process. Ultimately, this is the core idea of the actor-network.

As a first step Latour's method of reading can be understood as a comparatively simple semiological recipe for dealing with written sources from the history of science. Although at this point he cites the "structural semantics" of Greimas, which he had used in his collaboration with Fabbri, the reference remains general: "The semiotic method is here limited to the interdefinition of actors and the chains of translations" (PF 11). Or, put differently: "The method I use here consists simply in following all these translations, drifts, and diversions as they are made by the writers of the period" (PF 11).

In this connection, Latour refers to the entry "actor" in the semiotics dictionary that Greimas published with Joseph Courtés. However, Latour subverts his own attempt to define this concept by listing a series of synonymous expressions:

> I use "actor," "agent," or "actant" without making any assumptions about who they may be and what properties they are endowed with. Much more general than "character" or "dramatis persona," they have the key feature of being autonomous figures. Apart from this, they can be anything—individual ("Peter") or collective ("the crowd"), figurative (anthropomorphic or zoomorphic) or nonfigurative ("fate"). (PF 252)[11]

Greimas, however, does not mention translations, but this does not bother Latour because he is not seeking eventually to define the concepts "actor" or "actant" exactly. Although he frequently launches attempts at clarification, as in an essay co-authored with Madeleine Akrich on the sociology of technology, or in the appendices to his books *Politics of Nature* (PN 237–251) and *Pandora's Hope* (PH 303–312),[12] the definitions given there in glossaries are not consistent, they complement one another, and they overlap.

To make at least a small contribution to clarifying this further, it should be pointed out that when Greimas introduced and used "actor" and "actant," he said he was inspired by two works in particular: Vladimir Propp's *Morphology of the Folktale* (published in Russian in 1928)[13] and the philosopher and aesthetician Étienne Souriau's *Les deux cent mille situations dramatiques*—"The Two Hundred Thousand Dramatic Situations"—published in French in 1950.[14] Greimas saw these two works as both pursuing the same project: to organize literary "microuniverses" (deriving from folktales and theatrical works) by demonstrating the limited number of recurring "actants" with fixed characteristics, for example, the villain, the arbiter, or the savior.[15]

The twist here is that *actants* can be represented by various *actors*. The villain actant, for example, can be a lion or a stepmother. According to Greimas, psychoanalysis proceeds in a similar way: the psychoanalyst looks for actantial relationships (e.g., the oedipal triangle) in which positions or roles can be filled by specific actors (e.g., father or uncle). In this way some actants can be ascribed to several actors, but not all actants manifest themselves in actors (e.g., the weather).

This complex state of affairs is rendered even more complicated because Greimas not only refers to actants on the level of the content of entire texts but also on the formal level of individual textual components. Therefore, following Lucien Tesnière, whom he credits with introducing the linguistic concept of "actant," Greimas says that each and every sentence should be understood as a kind of theatrical stage upon which every component of the sentence (subject, verb, object, etc.) functions as an actant.[16]

Into these opaque depths of linguistics and semiotics Latour quite understandably does not venture. Instead, he returns to the procedures that he used in his sociosemiological analyses with Fabbri and Woolgar. Thus the first part of *The Pasteurization of France* offers a quantitative assessment of the various research themes featured in the *Annales* of the Institut Pasteur during this period. The statistics compiled from the titles of articles are presented in figures, such as curves and bar graphs. Similar to *Laboratory Life*, the original French edition of the book on Pasteur embeds the figures in the body of the text (MI 112, 114), whereas in the English translation they are relegated to the appendix (PF 268–269). In addition, a qualitative analysis is made of all the statements of authors published in the journals that relate to the changes in French society brought about by Pasteur's scientific findings. The results of this survey, which targets the corresponding changes in the "rhetoric" and "style" of the articles, are also presented by means of graphic methods (MI 78; PF 267).

The Politics of Knowledge

The concept of an actor-analysis cited above in which the analyst records "what each actor says of the others" should therefore not be understood as an empirical realization of Serres's philosophical agenda nor as a simple taking over of established semiological techniques. Rather, it is a creative draft for a semibacteriological, semiexegetic method in which the process of tradition and transmission becomes redefined.

As described above, in the first part of *The Pasteurization of France* Latour extends the epistemology into a kind of epidemiology. He conceives of the public demonstrations by Pasteur as intentionally infecting an entire society with microbiological knowledge. The laboratory in this context appears in the role of a "pathogen," an infectious agent that spreads via various paths of transmission and in the course of this changes its effects. When Latour says of his study, "I have spoken of the Pasteurians as they spoke of their microbes" (PF 148), he is confirming his endorsement of an epidemiological model of the diffusion of knowledge.

Yet when in the second part of his Pasteur book, the philosophical "Irreductions," he begins to explain the basic concepts of his theory again, Latour reiterates the view advanced in his earlier publications that science is an eminently polemical endeavor. The notion that scientific practice should be understood as agonistics, though, is not explained here with reference to Lyotard and Bourdieu but to Nietzsche—a Nietzsche, moreover, who is clearly seen through Deleuze-tinted glasses. As Karin Knorr-Cetina has pointed out, the "Irreductions" section contains the formulation of "Latour's Nietzschean theory of the political nature of all social life."[17]

In actual fact it is a Deleuze-influenced exposition of the distinction between knowledge and power, reason and force, cognition and politics. Here, too, Latour does not distinguish between the concepts of actor and actant but even adds as synonyms "entelechy," Leibniz's monad, and above all "force" (and its opposite, "weakness") in the sense used by Nietzsche, for example, as in "When one weakness [i.e., force] enlists others [other entelechies], it forms a network so long as it is able to retain the privilege of defining their association" (PF 170).

At the very beginning of his book on *Nietzsche and Philosophy* (first published in French in 1962), Deleuze points out the significance of the force concept: "We will never find the sense of something (of a human, a biological, or even a physical phenomenon) if we do not know the force which appropriates the thing, which exploits it, which takes possession of it or is expressed in it."[18] Additionally, Deleuze emphasizes that Nietzsche's concept of force is a force that refers to another existing force. Therefore, like in a network every force must be seen as a part of relations, which themselves are to be understood as relations of interpretation and domination. To put it more pointedly: all interpretation (*Ausdeutung*) is exploitation (*Ausbeutung*). The forces coexist and link up to one another. At the same time, they struggle to dominate one another in order to create differences.

As an exegete of science, for Latour it certainly is illuminating that Deleuze sees the full interpretation of the descent and the origin of the forces that act in these fields as a basic problem for engaging with modern science. According to Deleuze, every fact is an interpretation that in turn needs to be interpreted.[19] At least partially Latour's analysis points in this direction: Pasteur's public vaccination of animals at Pouilly-le-Fort literally dramatized scientific truth. And Latour in turn dramatizes the dramatization: Who is talking? Who is watching? And who confirms and authenticates? This puts the theater of proof onto the stage of actantial analysis.

However, it is open to question whether Nietzsche's conception of the creative and competitive interpreting forces can be reconciled with the world of networking actors described by Latour in the first part of his Pasteur book.

Pasteur's mise en scène of proofs can be understood as an extremely polemical, even hostile act. By contrast, the "interdefinitions" of the actors in the pages of the *Concours Médical* or the *Annales de l'Institut Pasteur* seem to follow the logic of enumeration and association rather than the will to power.

In David Bloor's view, the almost palpable tension in the air at this point can be attributed to the fact that it still remains completely unclear how to connect Latour's "metaphysical talk to historical and everyday reality."[20] The tension, though, is also caused by the difference between two biologies. Nietzsche had by no means developed his theory of force as a purely philosophical theory but in conjunction with an intensive dialogue with the life sciences of his period. These did not include bacteriology, as with Latour, but general physiology, Darwin's theory of evolution, and especially Wilhelm Roux's developmental mechanics, with its idea of the struggle between body parts within an organism.

Unfortunately, we do not learn anything about the scientific background to Nietzsche's philosophy from Latour, which seems all the more a regrettable omission because the author of *The Gay Science* functioned as an early ethnologist of the university and was also a philosophical visitor to laboratories.[21] For the time being there remain two separate discourses, or *dissoi logoi*: on the one hand Latour's associated actor-networks and on the other Nietzsche's differential relations of force.

Irreductionism

Apart from the propositions, demonstrations, and scholia, the philosophical part of *The Pasteurization of France* contains narrative passages similar to those found in other works by Latour. A total of eight of these "Interludes" are intercalated in the text, for example, to inform the reader about "Why This Treatise Says Nothing Favorable About Epistemology" and that "There Is No Such Thing as a Modern World." In the first interlude the author's aims are also illuminated in "Pseudoautobiographical Style." Put simply, these consist in developing "irreductionism," the main motto of which is: "We do not want to reduce anything to anything else" (PF 156). According to Latour's account, his first realization of this principle stemmed from an experience that could not be reduced to anything else.

Latour dates the event to the winter of 1972, when for a short time he taught philosophy in Gray. During a car drive from Dijon to Gray he had the overwhelming impression of suffering from an overdose of reductionism:

> A Christian loves a God who is capable of reducing the world to himself because he created it. A Catholic confines the world to the history of the Roman salvation. An astronomer looks for the origin of the universe by deducing its evolution from the Big Bang. . . . A philosopher hopes to find the radical foundation which makes all the rest epiphenomenal. (PF 162)

Latour was forced to stop the car. He parked by the side of the road and looked up into the trees. A few moments later he banished the avalanche of the reductionists by repeating to himself: "Nothing can be reduced to anything else, nothing can be deduced from anything else, everything may be allied to everything else" (PF 163).

This contraposition contains, as Graham Harman has noted, a kind of primal scene of Latourian philosophy.[22] Latour's more recent work as a metaphysics of plural "modes of existence" thus begins with a personal experience in 1972. In the book on Pasteur this primal scene stands in relative isolation, however. It is a marked contrast to Pasteur's demonstration described in the first part of the book with which he succeeded in convincing the farmers and vets of Pouilly-le-Fort. It would be better to compare Latour's story with how the idea of eternal recurrence suddenly came to Nietzsche by the famous stone near Surlej in Engadine during a walk around the lake.

One can also draw other, less spectacular connections, including to another, little-known "irreductionist" of the philosophy and sociology of science. It is not so much Marc Augé who comes to mind here. As we have seen, for Augé the logic of a social system is nonreducible to the logic of one of its component parts. At this point we are led—via Serres—to Auguste Comte.

Interlude with Comte

"Today we would need an Auguste Comte," writes Serres in the third volume of *Hermes*, which is devoted to the problem of translation. What is lacking at present is a classification of the sciences, or at least an idea of whether such a classification is still feasible. "The crisis of our knowledge has no location," complains Serres, and he opines that in this situation an Auguste Comte would at least provide a reference point.[23]

Thus it was entirely consistent that almost simultaneously with the volume of *Hermes* containing this diagnosis Serres published a new edition of Comte's most important work, *Cours de philosophie positive.* A few years later Latour confirmed this interest in Comte when in 1985 he prepared a new edition of Comte's *Traité philosophique d'astronomie populaire* at Serres's request. In this context positivism no longer appeared to be a philosophy of progress or a doctrine of universal scientific explanations in the service of the most rigorous economy of thought. Instead, Comte's work in Serres's view is a philosophy of cycles and recurrent comparisons: "Positive philosophy oscillates between invariants, it translates the same themes time and again. It is a cycle of translations, a repetitive encyclopedia."[24] Latour, the sociological theoretician of translation who sought to draw a new map for orientation within the history of science with his *Pasteurization* book, will have taken careful note of this.

In actual fact, it is possible to demonstrate that Comte's positivist philosophy can already be understood as irreductionism. Optical phenomena, for example, Comte does not regard as traceable to mechanical or acoustic facts and events. As he emphasizes in the *Cours*, "the phenomena of light will always represent a category *sui generis* that is necessarily nonreducible [*irréductible*] to all others."[25] Furthermore, Comte assumes that disciplines like biology will not be absorbed into chemistry or physics in the long term and that sociology will not vanish into biology. Comte regards these disciplines and their objects of study as possessing specificity, and in his view future developments in science will not change this fundamentally.[26] Because of the special role that this assigns to biology, Comte's philosophy also represents an important reference point for Canguilhem's "vital rationalism." Indeed, vitalism can be regarded as exemplary irreductionism; it is not least for this reason that Canguilhem invokes Comte as the founding father of historical epistemology.[27]

Serres has no objection to this. However, he gives a fresh slant to Comte's role when he introduces him primarily as a sociologist of religion and theorist of fetishism and not mainly as a historian of science. In 1989, in his *Elements of a History of Science*, Serres even goes so far as to present Comte as the founder of "the anthropology of the sciences."[28]

With this, one of the arenas is reached in which Latour's activities during the *Pandore* years will play out. As part of a group of philosophers, sociologists, and humanities scholars supervised by Serres, which included Bernadette Bensaude-Vincent, Geoffrey Bowker, Jean-Marc Drouin, Pierre Lévy, and Isabelle Stengers, Latour contributed to a comprehensive history of science whose most striking characteristics were that it was no longer positivist in the conventional sense, that is, oriented on the gradual, empirical accumulation of scientific facts, nor was it in any way operating under the banner of the epistemological break or the sudden shift of paradigms.

Serres formulated the program of this historiography as the science history of networks:

> Far from tracing a linear development of continuous and cumulative knowledge or a sequence of sudden turning-points, discoveries, inventions, and revolutions plunging a suddenly outmoded past instantly into oblivion, the history of science runs backwards and forwards over a complex network of paths which overlap and cross, forming nodes, peaks and crossroads, interchanges which bifurcate into two or several roads.[29]

This is the exact conception of a network that *Science in Action*, Latour's next book, will adopt with respect to the sociology of science. In fact it had already been published when Serres presented the program cited above.

A History of Things

Latour's own program for a historiography of science has a different agenda. It aims at a *history of things*. "Things" are not meant here in an everyday sense; Latour is not concerned with material objects like a table one can sit at or a laboratory bench on which one sets up apparatus. Nor is it about the carefully assembled things found at flea markets, which would allow historical materialists to be led in their investigations "by the objects themselves," as Walter Benjamin put it.[30] The things Latour is interested in are specific to science: the facts or topics that scientists engage with and those realities that they term as "new" and that they claim to have "found." Most of the historical sources that Latour discusses in this context stem from his study of Pasteur. One example is his discussion of the debate between Pasteur and Pouchet on spontaneous generation of organisms and the "discovery" that lactic acid bacteria are responsible for fermentation processes.[31]

The starting point of his historiographical contributions is what Latour terms the "anthropological matrix." This matrix demonstrates that collectives permanently self-renew by organizing themselves around things that are themselves in a state of permanent renewal (NB 85). Thus the scientific object takes the place of Serres's quasi-object. It holds a laboratory collective together like a ball keeps the players of a football or rugby team together on the pitch. In consequence, Pasteur's microbe is not an entity that was always a given but "the provisional form of the networks" that are requisite for the composition and definition of this "thing." Thus the object originates from out of a multitude of crisscrossing aspects and tendencies:

> a friend of the Emperor's, a tool of microbiology, a response to Liebig, destroyed by the heat, transportable by air and by clothing, stopped by the curves in the glass, shatterer of atheism, parent and child of creatures exactly resembling it, anaerobic respirator, the promise of a solution to life, death and illness, absent from glaciers, present in Paris, mastered in the rue d'Ulm, this is how it appears in the heat of the controversy [with Pouchet], as far as Pasteur is concerned.[32]

From an actantial analysis—like in the *Pasteurization* book—that merely records "what each actor says of the others," we seem to have come a long way. In fact, the criterion that Latour invokes with regard to the "thingness" of lactic acid bacteria and microbes is no longer linguistic but pragmatist. It is not decisive anymore what the actors *say* about one another but what they *do* with and against one another.

In this sense Latour explains that the microbe *is* nothing more than the "list of actions and of tests" that were performed in Pasteur's laboratory.[33] Or, similarly: "Actors gain their definition through the very trials of the experiment" (FR 66). And finally: "An actant is a list of answers to trials—a list which, once stabilized, is hooked to a name of a thing and to a substance" (TD 122). In other words, the list of tests and trials *is* de facto the history of the entity that was subjected to them.

As a key witness for this conception of a history of things Latour calls upon the pragmatist William James, who construed existence primarily in terms of action. Whereas James had applied this above all to "thinking beings," Latour extends it to include "nonhuman actors who have also their history" (FR 78–79).

A few years later, Isabelle Stengers will suggest to Latour that he should draw on the philosophy of Alfred North Whitehead for his renewed interrogation of Pasteur with a view to characterizing the specificity of scientific objects more precisely. From the perspective of Whitehead's philosophy, which is also pragmatist but less centered on the human subject—Latour speaks of "realism without substance"[34]—the lactic acid bacterium becomes a peculiar thing whose "competences" can only be defined by taking onto account its "performances."[35] At the same time this "discovery–invention–construction" is an induced event which affects and modifies all the relations between the actors involved—both microbiology and the microbiologists themselves.[36] Thus already at the end of the 1980s Latour was starting to define actors through their agency. Once this was extended to include technical objects, it proved to be one of the most controversial postulates of Latourian anthropology.

SIX

Science and Action

Around 1985 the Center for the Sociology of Innovation at the Écoles des Mines in Paris became the institutional basis of actor-network theory. Although "theory of the actor-network," "actor-network theory," or simply "ANT" did not become common currency until the early 1990s,[1] the "author network" of actor-network theory had already formed several years previously at the Center for the Sociology of Innovation. The association between Latour and Callon was the nucleus of this network. Callon, Lucien Karpik's successor, was director of the center from 1982 to 1994. Latour also arrived there in 1982 and stayed for almost a quarter of a century. During this period he wrote several of his most famous works, including *Science in Action* and *We Have Never Been Modern*. Together with Callon Latour also wrote a number of programmatic essays: for example, the introduction to the anthology *La science telle qu'elle se fait* and "Unscrewing the Big Leviathan."[2] In the French original of *The Pasteurization of France*

Latour expresses his thanks to Callon, "who in many respects is the co-author of this book" (MI 7). *Science in Action*, which was published a short time later, is dedicated expressly to Callon as an "outcome of a seven-year discussion" (SIA VII).

From the mid-1980s the Latour-Callon nexus was supplemented by several further associates; of particular note were the external collaborations with Greimas's student Françoise Bastide in Paris, the sociologist John Law in England, and Arie Rip in the Netherlands, who in turn had their own international associates, for example, Annemarie Mol and Susan Leigh Star.[3] In the years that followed, the human resource base of ANT was expanded at the Center for the Sociology of Innovation, for example, by Madeleine Akrich, a sociologist of technology, and the music sociologist Antoine Hennion; both scholars went on to become directors of the center. The year 1985 also marks the beginning of efforts to propagate the sociological approach cultivated at the École des Mines as distinct and independent. An anthology, for example, which Callon edited with Law and Rip, included a glossary of key concepts, such as "actor-network," "black box," "enrollment," and "translation centers," with brief definitions and explanatory notes. Additionally, the volume began with a programmatic chapter ("How to Study the Force of Science") and provided information about the central elements of the ANT methodology ("Follow the Actors"). The main issue at that time was "qualitative scientometrics," an attempt to measure and analyze science with a view to mapping dynamic processes in science. It is worth noting that in this connection "actor-network" meant a network that is itself an actor—a fractal understanding of the two terms that is not often found in Latour in this form.[4]

An Anthropology of Science

Published in 1987, *Science in Action* can be seen as Latour's contribution to the beginning consolidation of ANT. Its subtitle proclaims that it is a textbook, a manual, and a proper compendium: *How to Follow Scientists and Engineers Through Society*. After his *Pasteurization* foray into literature and philosophy alongside Tolstoy and Nietzsche, with *Science in Action* Latour makes use of a genre that, similar to the text of *Laboratory Life*, which is riddled with tables and diagrams, can be fairly easily put into isomorphism with similar genres in the sciences. Here, too, Latour includes in the introduction references to literature, this time in terms of recommending publications by Tom Wolfe and Tracy Kidder, which as he puts it, have written two of the "best books" on technoscience and are therefore "compulsory reading . . . for all of those interested in science in the making" (SIA 260).[5] However, the complicated text layout, with its many instances of italics and boldface, set-off paragraphs, and numbered examples, and particularly the number of figures and even a few cartoons, is above all highly reminiscent of biology and physics textbooks for beginners. Originally, Latour even planned to add some exercises for the reader to do at the end of each chapter (SIA 17).

The main purpose of *Science in Action* is a trip through the unfamiliar territory of "technoscience." This term, which was introduced in a French-speaking context by Lyotard, among others, in connection with his theory of postmodern knowledge,[6] in Latour's understanding signals his ambition to be not only interested in the sociology of science but also in the sociology of technology. And in fact *Science in Action* does undertake to develop a balanced perspective toward the construction of "facts" *and* "machines," of "scientific facts" *and* "technical artifacts."

This not only takes into account the "the technical criterion, introduced on a massive scale into scientific knowledge," which according to Lyotard has led to an ever greater orientation on epistemic performance in scientific practice.[7] Rather, in this way it should also be possible to place scientific objects, like the DNA double helix, on a level with the production and distribution of minicomputers and other technical objects (SIA 3). For Latour both cases are "black boxes" that can be opened and studied with regard to content, origin, and distribution.

In this connection, one of the functions of the term "technoscience" is that it sets up a parallel between "fact" and "machine." Further, it serves as a kind of test probe "to describe all the elements tied to the scientific contents no matter how dirty, unexpected, or foreign they seem" (SIA 174). This is a less ambitious but perhaps more plausible version of the term. Understood in this sense technoscience appears to be merely a synonym for "*science* in context"—which also reflects the fact that up to this point in time Latour had not conducted any independent technosociological research of his own.

In keeping with the character of a textbook *Science in Action* mainly addresses first-year students who are seeking thematic and methodological orientation in the field of science and technology studies. For these freshmen, whom Latour frequently addresses directly, the appendix provides guidelines such as: "Rule 1: We study science *in action* and not ready-made science or technology; to do so, we either arrive before the facts and machines are blackboxed or we follow the controversies that reopen them" (SIA 258).

But the novices are not just any kind of naïve students. In Latour's eyes, their character is quite special; they are skeptical and mistrustful of the statements made by scientists and engineers. Latour describes these student figures also as "dissenters" and "doubters." They appear partly like a police detective who wants to investigate the scientists and their seemingly secret activities and partly like a latter-day doubting Thomas who cannot overcome his doubts about belief in science as long as he doesn't see what the scientists are doing with his own eyes or feel it with his own hands. Here Latour recruits again the position of the agnostic, which we have encountered already. *Science in Action* unfolds as an often humorous, at times strenuous dialogue with these unbelievers who are interested in what seems to function as the last religion: science.

The result is again a sophistic image of science. Although this image does not separate into two views, *dissoi logoi*, as in *The Pasteurization of France*, which would correspond roughly to an empirical and a theoretical part of the book, the dissenters still have to contend all the time with two different views on their trip through technoscience. On the one side there is ready-made science and on the other science in action (SIA 13); on the one hand nature as the *cause* that allows controversies to be resolved and on the other nature as the *result* of the settling of controversies (SIA 99); on the one side science and technology as reasons *behind* the realization of actual projects that, on the other side, point *toward* science and technology as reasons (SIA 175).

Latour speaks in this connection of technoscience as a two-faced Janus, and there are indeed many drawings of two-faced figures inserted in the text. But this is not a charge of ambivalence—on the contrary: Latour is concerned with a symmetrical perspective, a "binocular view" of science, that allows both its past and its future to be visible, its inside and outside, the text and the context. For Thomas Kuhn reversible figures, which can be seen as one thing or another (like the famous vase or two faces), were a model borrowed from gestalt psychology for the paradigm shift, which temporarily suspends the smooth continual progression of science.[8] Latour strives to consider both views of science at the same time and on this basis arrive at "a stereophonic rendering of fact-making" (SIA 100).

In the Hinterland of the Texts

Given the suggestion in its title, the big surprise of *Science in Action* is that it starts by postponing the encounter with scientific practice for the time being. Unlike his earlier work with Woolgar, Latour does not begin by visiting a laboratory (or a workshop) but commences by taking a concentrated look at "the literature"; that is, articles published in peer-reviewed science journals, "the most important and the least studied of all rhetorical vehicles" (SIA 31). Here Latour not only returns to the examples he had already dealt with in his work at the Salk Institute—the publications on neuroendocrinology by Guillemin and Schally—he also reemphasizes the polemical, agonistic character of scientific work and returns to the phenomena of positive and negative modalities of statements, which he had already addressed with Fabbri in his study of the rhetoric of science.

Latour underlines here once again that the special character of scientific texts consists in omitting modalities in statements as far as possible. He also draws attention to the genuinely exegetic element of this process: "*the status of a statement depends on later statements*" (SIA 27). Thus positive and negative modalities are not only phenomena that can be observed within the synchrony of a text; rather, they continue in the diachronic cross-referencing of texts. This is a decisive aspect for Latour and of such significance that he admits that *Science in Action* is simply "a development of this essential point" (SIA 27). In other words, that which will be the past one day in the future determines the present of scientific texts and contexts.

The second chapter then leads into the laboratory; in a "showdown" Latour proceeds "from texts to things" (SIA 64–79). Or at least this is how it seems, because the laboratory chapter is introduced according to a bold formula: the laboratory director explains to a dissenter how a certain body substance, which is mentioned in a published article, can be traced by means of an inscription device. The director takes the role of a "spokesman" or "mouthpiece" of the substance in question: "See? . . . Here is endorphin" (SIA 71), he says to the visitor, who is looking at the graphs the director is pointing at with his finger. Latour describes this as the quasi-political function of the scientist, who here acts as the representative of the substance in question. In this way he substantiates the reflections on the problem of translation as per Callon and prepares the ground for the "parliament of things," which he will elaborate at a later point in time.

For Latour the process of representing is essentially based on an inscription; in the aforementioned example, it is on a curve displayed on a laboratory instrument called a "physiograph."[9] In addition the process is an indication of a "trial of strength" in which the association of the actors involved is tested for its stability and then affirmed. Accordingly the process of representing is described in the following terms:

> What is behind the claims [of the scientists]? Texts. And behind the texts? More texts, becoming more and more technical because they bring in more and more papers. Behind these articles? Graphs, inscriptions, labels, tables, maps, arrayed in tiers. Behind these inscriptions? Instruments, whatever their shape, age, and cost that end up scribbling, registering, and jotting down various traces. Behind the instruments? Mouthpieces of all sorts and manners commenting on the graphs and "simply" saying what they mean. Behind them? Arrays of instruments. Behind those? Trials of strength to evaluate the resistance of the ties that link the representatives to what they speak for. (SIA 79)

Thus the showdown does not simply proceed from the texts to the things but from the published papers to the instruments or inscription devices. In fact, for Latour "instrument" and "inscription device" are virtually synonymous: "I will call an instrument (or inscription device) any set-up, no matter what its size, nature, and cost, that provides a visual display of any sort in a scientific text" (SIA 68). As a consequence, it is again not the materiality of the instrument that is the decisive factor nor any output in acoustic signals (e.g., as in radio astronomy). Latour's concept of an instrument rests on the visual relationship of signs to what is designated, that is, the graphic reference: "The instrument, whatever its nature, is what leads you from the paper to what supports the paper" (SIA 69). This can be the chain of notes and images, which in soil research leads from the earth in a particular region to a portable map. As we shall see, it can also be a large sailing boat on a mission to map an unknown island.

Great Divides, Large Networks

The first two sections of *Science in Action* remain in the relatively familiar world of the sociology and history of science, but the third enters the largely unexplored terrain of anthropology. The two-faced Janus of on the one side ready-made science and on the other science in the making, is extended to the pairings knowledge and belief, the universal and the local, the existence of Westerners ("we") and of non-Westerners ("they") that in themselves now appear as reversible figures.

In his study *The Domestication of the Savage Mind* (1977) the British anthropologist Jack Goody had spoken of the "Great Divide" to challenge the problematical dichotomization of categories in public and academic discourse regarding the relationship of "advanced" and "primitive" civilizations: science/mythology, engineer/bricoleur, thought/perception, modern/ Neolithic, and so on.[10] At one and the same time Goody had also demonstrated that there are a number of cultural techniques that circumvent or cross the Great Divide, for example, making lists or following recipe-type prescriptions.

Following Goody, Latour states that it is the phenomenon of inscription that allows to develop a balanced anthropological view of science and technology, and he illustrates this by the historical example of the exploration frigates *Astrolabe* and *Boussole*. Jean-François de La Pérouse, the geologist and explorer, was commissioned to map the eastern Pacific with these ships in the 1780s by Louis XVI. By taking this particular example Latour is able to bring into play the motifs of navigation and orientation, which Serres is so fond of, as well as the cybernetic theme of the art of the helmsman.

At the same time Latour links the phenomenon of inscription to that of the network. By now inscription for him has become much more than a strictly local phenomenon, as in *Laboratory Life*; it now stretches beyond the confines of individual laboratory facilities and takes in very different people and institutions. Thereby cartography becomes the paradigm of knowledge whose central characteristic is the familiarity with events, locations, and people (SIA 220) while at the same time being able to *dominate* them from faraway places (SIA 224).

Instead of foregrounding the dynamic interaction of competing forces, which within research institutes form associations and differentiate themselves through reciprocal translations, Latour now uses the term "network" to evoke the notion of overarching technological infrastructures:

> The word network indicates that resources are concentrated in a few places—the knots and the nodes—which are connected with one another—the links and the mesh: these connections transform the scattered resources into a net that may seem to extend everywhere. (SIA 180)

The prototype of this is the railway network, which as we know also fascinated Péguy. Maps of itineraries and the routes of expeditions, communication systems like the telephone, and international measurement standards are other examples mentioned by Latour, who clearly has the "large technical systems" in mind that the historian of technology Thomas P. Hughes has described in an exemplary manner for the electrification of North America.[11]

But unlike Hughes Latour is not interested in electrical grids but in networks of signs, in semiotic infrastructures. How does one transport an inscription from A to B? With the example of the scientific article in mind, the answer would be: By printing the inscription on paper where it is surrounded by written text, binding it together with other pages, and sending it by post. And what if there is no post? Then one gives the printed paper to a traveling salesman, for example, who regularly makes the journey from A to B. And what if there is no traveling salesman? Then one has to find a map and set out with the paper oneself. And what if there is no map? Then—as in the case of La Pérouse—one has to get hold of a ship equipped with navigation instruments, telescopes, and drawing implements to make the requisite map. In other words, scientists' activities are not confined to the laboratory; their writing activities reach far beyond the bounds of these spaces of knowledge. Within networks of all sorts scientists make traces "circulate better by increasing their mobility, their speed, their reliability, their ability to combine with one another" (SIA 232).

According to Latour, this also distinguishes the activity of engineers. Their objective is to manufacture technical objects that can be used inside or along certain infrastructures, for example, trains, televisions, or computers. Thus the products of scientific and technological practice can be characterized, in general, as follows:

> Facts and machines are like trains, electricity, packages of computer bytes or frozen vegetables: they can go everywhere as long as the track along which they travel is not interrupted in the slightest. (SIA 250)

From the "Immutable Mobiles" to the "Centers of Calculation"

Networks, which are necessary for disseminating "standards and metrology" (RS 228), are not the only decisive factor for sending traces. To be transported safely, the traces themselves need to have a certain form and quality. If papers are to be sent by post, they should be well-protected or relatively robust objects but not heavy. If they include figures, these should have a high-enough resolution, while their exterior dimensions should remain standard. The same applies to sending seeds, soil samples, animal specimens, and other semiotic things. They, too, must be as mobile, stable, and combinable as possible, in order that they may be used productively for "domination at a distance" (SIA 223).

It is in this connection that Latour introduces the concept of "immutable and combinable mobiles." The concept denominates "charts, tables, and trajectories" that are "conveniently at hand and combinable at will" (SIA 227). Like the stuff found on a desk in a laboratory, these are typically flat objects: "At one point or another, they [the immutable mobiles] all take the shape of a flat surface of paper that can be archived, pinned on a wall, and combined with others" (SIA 227). Besides printed texts, other examples are lists, questionnaires, and diagrams (SIA 238). The production, archiving, and utilization of such "immutable and combinable mobiles," however, is only one aspect of the "paper world" (SIA 226) in which scientists and engineers operate. According to Latour, it is also a matter of processing the traces and signs in question at certain points in the network and feeding them back into circulation in a correspondingly modified form.

This is the task of the "centers of calculation." By this term Latour does not primarily mean the data centers as run by big companies, administration departments, and universities. Rather, what Latour has in mind are astronomical observatories, natural history museums, geological surveys, and official census bureaus (SIA 233). These institutions both collect stable and mobile traces of all kinds—from sediment samples to questionnaires and numerical data—and process them, use them as the basis for calculations, and put the results in a form that will guarantee simple and long-term subsequent use. Latour sees centers of calculation of this kind as the real heart of scientific and technological networks: "They tell us what is associated with what; they define the nature of the relation; finally, they often express a measure of the resistance of each association to disruption" (SIA 240).

The activity on which this is based is the catchy phrase "drawing together" (SIA 225, 239); drawing both in the sense of "pulling" and in the sense of producing an image or a diagram by making marks on paper. It is from the instruments and the institutions, which in this sense permit the association and combination of relevant knowledge elements, that an "anthropology of science" must start, in Latour's view, if it wishes to bridge the Great Divide between "us" and "them," between the domesticated and the savages: "In other words, the logistics of immutable mobiles is what we have to admire and study, not the seemingly miraculous supplement of force gained by scientists thinking hard in their offices" (SIA 237). The program Latour outlines here is still topical today, for example, in the area where history of science and media studies overlap.

Media Studies

Science in Action marks the end of Latour's Pandora years. The book theorizes the socio- and technosemiotic process of "drawing things together"[12] and, moreover, it practices this, too. *The Pasteurization of France* had to contend with being divided into two, into an empirical and a philosophical perspective. By contrast *Science in Action* is all of a piece, an amalgamation of exegesis and ethnology in an empirical philosophy of knowledge and belief in a world that is increasingly interconnected via technology. In this respect the book itself can be described as an artful network, an alluring machine. At a point in time when the envisaged renewal of the sociology of science and technology could only rely on approaches that were sparse and dispersed, Latour establishes an approach that is largely a coherent whole. An approach in which, based on numerous examples, the natural and the artificial, the earthly and the divine, the Western/Modern and the non-Western/nonmodern are drawn together in a stereoscopic view.

The name of this emerging mixture is, naturally, not Pandora but "anthropology of science." This project, which began in the early 1980s, Latour understands as directed against "the complete asymmetry between science and other systems of belief" (CR 208). He admits, however, that he uses "anthropology" in this connection as a "metaphor" (CR 206). Evidently, Latour's focus is not on foregrounding humankind as the subject but on the descriptive methods of ethnography, although the details of this methodology are often left in the dark. He also speaks of anthropology "with a dash of provocation" (CR 206). Wasn't it the specious cleaving to the human subject as concrete point of reference and reflection that structuralism sought to challenge? Not least in order to sharpen his own profile Latour is quite happy to provide a contrast.

Moreover, the label "anthropology" allows him to make connections between yesterday and today, the historical and the sociological, which are not only philosophical (as in Comte) but also empirical. Only the anthropologist's perspective reveals the ancient process of inscribing and interpreting traces to be of seminal importance for scientific and technical rationality, and not only for so-called primitive societies.

Through its reference to visualization Latour's approach acquires considerable suggestive power. Despite that its contours have scarcely been outlined, the anthropology of science already starts to open itself up to linkages with art history and media studies as well as to the emerging field of so-called image studies, or *Bildwissenschaft*. Obviously, the subject of visualization was not a new theme in the 1980s. The role of perceptions, images, and visual thinking in a science context was prevalent in France long before Latour. In the 1950s, for example, Alexandre Koyré had pointed out the significance of visual techniques for rational argument in his study of Erwin Panofsky's book on Galileo. Derrida's early translation and extensive comments on Husserl's text regarding the origins of geometry should also be mentioned in this connection.[13]

In an environment that in the 1970s and 1980s was marked first and foremost by an interest in linguistic structuralism, there had been another author who had privileged the visual sphere as opposed to that of the text, an author who was no stranger to Latour. Jean-François Lyotard, in his *Discours, Figure* (1971), had advanced figural representation as a subversive dimension with active lines and multidimensional points against the logic of language and thus, in Deleuze and Guattari's view, had provided "the first generalized critique of the signifier."[14] Although Lyotard's argument is aimed at the area of art and desire, he also discusses authors such as Bachelard, Panofsky, and Koyré and devotes entire passages of his book to the relationship of science and myth. So when Latour turned to "visual culture" in the early 1980s,[15] from today's perspective it might seem as though he had anticipated the so-called pictorial or iconic turn that was proclaimed by W. J. T. Mitchell, Gottfried Boehm, and other historians of art. Actually, though, Latour does not make any turn; rather, he synthesizes and accentuates already existing approaches and philosophical themes with which he was familiar.

Published in 1985, his article "Les 'vues' de l'esprit" provides the most comprehensively documented (and illustrated) version of the argument. Besides Koyré and Derrida Latour quotes Dagognet and Foucault. He also makes reference to Samuel Edgerton's texts on perspective and Williams M. Ivins on visual communication, which are both relevant for media studies. Latour refers particularly to the studies by Elizabeth L. Eisenstein on the printing press as an "agent" of scientific and religious change in the early modern period (VE 17). Some years previously in a review Latour had spoken favorably of Eisenstein's book, in which the changes in the cultural practice of Bible reading caused by the advent of printing feature prominently as a "most original materialism of inscription."[16] Indeed, his discussion of the immutable and combinable mobiles appears to be very much indebted to Eisenstein's examples of printed maps and books—including printing as such with its mixture of "typographical fixity" and "movable type."[17]

In this connection Latour names several research themes for the future in his planned anthropology of science. He proposes that more attention should be paid to the visual representation and calculation of machines and suggests pursuing an "ethnography of filing folders," a project that he recently tackled within the framework of being a participating observer of the Conseil d'État (ML 70–106). Yet at the same time Latour looks back; he quotes Bultmann and Péguy on the question of the trustworthiness of translations and makes a rare reference to his own work on "Exegesis and Ontology" (CR 232–234). And this is not the only instance when his anthropology of science turns back to earlier texts and contexts.

In the same connection an encounter takes place between the curves of Marey and the projective geometry of Monge. In "Les 'vues' de l'esprit" he refers to both of these scientists from Beaune; the former as an example of the director of a physiological "center of calculation" (VE 21) and the latter as a pioneer of the graphic representation of technical objects (VE 24). One might be tempted to quote Shakespeare that "the wheel is come full circle"—back to the city in Burgundy—if the circle was an appropriate metaphor here.

SEVEN

Questions Concerning Technology

A complete change of scene: Berlin. It is the ninth of November 1989. A press conference is being held. At the front of the podium representatives of the government have taken their seats at a long table. In the middle is the "Secretary for Information"; the audience is the usual assortment of journalists.

For an hour the press conference speakers drone on about issues of economic and domestic policy. Then the question of travel regulations comes up once again, a point that had already been resolved. The government spokesperson looks tired and a bit irritable.

The exegesis begins: the communiqué on this subject issued by the ministry concerned has already been handed out, he says. Nevertheless, on the podium there is whispering, papers are brought out, glasses put back on, and texts read over again. Then a reporter asks, "When will the regulations come into effect?" The government spokesperson shrugs his shoulders, his eyes wander over the text in front of him again. The translation follows: "As far as I know . . . effective immediately, without delay."

Shortly afterward the press conference ends, and the "Secretary for Information" drives back to the Central Committee building, collects his briefcase, and finishes work for the day. Even as he is driving home the run begins on the "obligatory passage points," which people have to pass through to get from East to West Berlin.

What later becomes known as the Fall of the Berlin Wall and a key event in the collapse of the Eastern bloc takes shape during the night. Alluding to Péguy, one might say that the people did not storm the Wall on a kind of whim but as the continuation of a constant movement of exegeses, rereadings, and revisions.

Naturally, Latour was not present at the press conference moderated by Günter Schabowski, nor did he ever comment on it. A few months before this event, in March 1989, he had visited West Berlin for the first time to give a lecture at the invitation of Bernward Joerges, a Berlin-based professor for the sociology of technology. At the Social Science Research Center Berlin, an independent research institution not affiliated with a university, Joerges had just completed a project on "Technology in Everyday Life" and was starting a new one on "Large Technical Systems."[1] It was within this context that Latour was invited to present his program for a symmetrical sociology of human and nonhuman actors.

A few months before, Joerges and Latour had met at the Annual Meeting of the Society for Social Studies of Science (4S) in Amsterdam, and they had immediately gotten on well together. "I am German, but I dare to like nonhumans," was how Joerges had introduced himself, pointing out that they had interests in common.[2]

In effect, both were interested in the social realities created by and through technical objects, a theme that in light of the expanding cable television and computer networks inevitably suggested itself to social and political scientists at this time, from Sherry Turkle and Wiebe Bijker to Langdon Winner and Hubert Dreyfus. Donna Haraway had aptly summarized this obtrusiveness of technology in her famous *Cyborg Manifesto* in the mid-1980s: "Our machines are disturbingly lively, and we ourselves frighteningly inert."[3]

Latour's Berlin trip took on a special significance because of his encounter with a thing. The object was on Joerges's key ring. As an inhabitant of Paris Latour was well acquainted with the small keypads and combinations of keyless entry locks by which one enters many houses in that city. However, a key like Joerges's he had never seen before—along with most everyone else on the planet.

At that time many tenements in West Berlin did not have electric door openers, and the main door of the buildings had a special locking system. To ensure that the tenants could come and go as they pleased but would lock the door at night, a key with two sets of teeth, one at each end, was inserted into the keyhole from one side, turned to open the door, and pushed through to the other side. After going out or coming in, one could only withdraw the key from the keyhole after turning it again, this time to bolt the door.

Latour was fascinated. He showered Joerges with questions, went home with him, tried out the key, and then went to a Berlin locksmith to buy such a lock and key to take home with him. Back in Paris Latour wrote a case study on "The Berlin Key," in which this technical object is revealed as the decisive agent mediating the social relations of the tenants of, users of, and visitors to a building (CB 33–46).

The full symbolic ramifications of the Berlin key, however, only became clear after the famous press conference of November 1989. From that point onward the surreal-looking key referenced the opening of the border between the two German states and the beginning activities of the "Wallpeckers." As a material object, as a key-thing, it also foregrounds the role that real technical objects played in this process. There is no doubt that reunification was above all a political and social event. However, the technological and media infrastructures, particularly radio and television, had a major share in it. The key thus opened up the works of Latour to contemporary politics. One could also say that at this time Berlin becomes a key to Latour's work.

Indeed, central motifs of the Latourian oeuvre—looking at assemblies of heterogeneous elements, the stereoscopic suspension of divisions, the bridging of discontinuities—take on extraordinary topicality through the fall of the Berlin Wall. Latour does not pass up this opportunity. He turns the Berlin key into the leitmotif of an anthology published in French in 1993, *La clef de Berlin*.[4] Before this, in *We Have Never Been Modern* (the German edition contains a dedication to Joerges), he relates themes such as symmetry and continuity to "1989, the year of miracles" and asserts there is "perfect symmetry between the dismantling of the wall of shame and the end of limitless Nature" (NM 17).

Ultimately, it is this double symmetry that the double-ended key embodies. The question is, though, how far one can go in ascribing to such "nonhuman actors" a momentum of their own and perhaps even agency. This is the problem that Latour's works engage with in the 1990s.

The Exegesis of Modernity

In *We Have Never Been Modern*, clearly his most successful book but only at first glance his most accessible, Latour characterizes modernity via the interaction of "two sets of entirely different practices" (NB 10). The first is "translation," which leads to the emergence of hybrid networks and objects, to mixtures of nature and culture. The second set of practices is "purification." It produces two completely separate ontological domains—humans on the one side and nonhumans on the other (NB 10–11). The decisive issue is not to offer an exact description of the two practices in question. In the first instance Latour is interested in bringing together the *dissoi logoi* in a single aspect, a stereoscopic view:

> As soon as we direct our attention simultaneously to the work of purification and the work of hybridization, we immediately stop being wholly modern and our future begins to change. At the same time we stop having been modern, because we become retrospectively aware that the two sets of practices have always already been at work in the historical period that is ending. Our past begins to change. (NB 11)

Thus initially it is about changing a viewing angle, a change that looks at the scheme of modernity in a whole new light (NB 11): above the Wall between East and West (purification), below the technological and media reunification (translation).

Since Latour had already described a modernity in *The Pasteurization of France* that "was not modern," one might suppose that he would develop his argument about the "constitution of modernity" again with reference to

Pasteur. This, however, is not the case. Instead, Latour takes a study by two historians of science, Steven Shapin and Simon Schaffer. In their highly influential *Leviathan and the Air-Pump* (1985), Shapin and Schaffer examine in parallel the political philosophy of Thomas Hobbes and the scientific practice of Robert Boyle.[5]

In Latour's opinion the study allows one to follow in an exemplary way the separation of the regimes of "political representation" and "scientific representation" (NB 27), a separation that is characteristic of the modern age (NM 40). Since the seventeenth century rather clear-cut distinctions have been made between the "politics of men" and the "science of things" (NB 30). On the one hand modernity cultivates the view that "human beings, and only human beings, are the ones who construct society and freely determine their own destiny" (NB 30), whereas on the other hand it follows the maxim that "it is not men who make Nature. Nature has always existed and always already been there" (NB 30).

Following Shapin and Schaffer, Latour depicts these two separate yet reciprocally referencing regimes of representation as different forms of exegesis. In his view it is not coincidental that Hobbes's main work, *Leviathan or The Matter, Forme and Power of a Common Wealth Ecclesiasticall and Civil* (1651), consists largely of an exegesis of the Old and New Testaments that exhibits features of a mathematical proof. Hobbes argues against interpretations of the Bible and at the same time against assigning agency to matter: "Inert and mechanical matter is as essential to civil peace as a purely symbolic interpretation of the Bible" (NB 19).

By contrast Boyle develops a new kind of text that adapts the old repertoire of Bible exegesis (and of penal law) in order to allow new kinds of entities to get a hearing: the nonhuman actors of the laboratory. As a consequence, Boyle's new type of text, the experimental science article, appears as a "hybrid between the age-old style of biblical exegesis—which has previously been applied only to the Scriptures and classical texts—and the new instrument that produces new inscriptions" (NB 23–24).

This is the second characteristic element of Latour's reading of Shapin and Schaffer's *Leviathan and the Air-Pump*. The first consists in the fact that he shifts the entire discussion of the work of Hobbes and Boyle from the relatively specific terrain of social history to the more general discussion of modernity that began with Comte and Péguy and achieved new relevance in the early 1990s with the sociological contributions of Ulrich Beck, Zygmunt Bauman, and Anthony Giddens.[6] In Shapin and Schaffer's book, however, the discourse on modernity is not an issue. A formative influence on the two historians was the Edinburgh school of the sociology of science, and in *Leviathan* they operate within the framework of social history that is oriented on authors such as Edgar Zilsel and—with respect to Hobbes—C. B. Macpherson. They see themselves as sociologists, not anthropologists, and whereas Latour is beginning to champion the "nonhuman actors," Schaffer and Shapin continue to uphold their perspective in which knowledge is "the product of *human* actions."[7]

Precisely this point is challenged by the second characteristic element of Latour's reading of *Leviathan and the Air-Pump*, which consists in making the air pump a "nonhuman actor," thus endowing it with a life of its own that it does not have at all for Shapin and Schaffer. For them the air pump is an instrument made and used by humans, a material component of scientific practice. For Latour, by contrast, the pump is an "actor," which, although it has no will and no soul, is "even more reliable than ordinary mortals to whom will is attributed but who lack the capacity to indicate phenomena in a reliable way" (NB 23). For this very reason, on account of its semiotic capabilities, it is possible for the air pump and the vacuum it produces to become the starting point of a new form of exegesis.

One can read this as Latour's rejoinder to Schaffer's critique of the "hylozoism" of his *Pasteurization* book, but at the same time it is an indication of the increasing importance of technical objects within Latour's work.

The Turn to Technology

In the foreground of Latour's sociology of technology case studies of the early 1990s is the everyday use of technology, which he analyzes by means of observations full of meticulous detail and a great deal of humor. Some passages evoke the early Alain Resnais' thing documentaries like *Le chant du styrène*, and others Charley Bowers's slapstick comedies featuring complex gadgets that have a life of their own. It is between these two poles that the studies contained in the *La clef de Berlin* anthology oscillate, which feature the symmetrical door opener, the door, the safety belt, the hotel key fob, and the automatic door closer. In other publications Latour gives similar treatment to the way people use cameras, video recorders, and speed bumps as well as commenting on the role of firearm possession in the United States.[8]

All of these case studies are informed by the attempt "to abandon the false symmetry between humans and things that confront each other" (CB 21). Instead, say Latour and Madeleine Akrich, the appropriate starting point for a sociology of technology will be "settings" or assemblies of humans and nonhuman actants among whom certain "competences and performances" are distributed.[9] Proceeding from such assemblages, the goal is to draw "a map of the alliances and the changes in alliances" (CB 19) that come about between human and nonhuman actors but that also disappear again.

Such maps are drawn using "sociotechnical graphs" as the basis. Horizontally these graphs give the changing composition of the respective setting ("association"), and vertically they display the changing "programs of action" in their sequence ("substitution"). The precept of the exegesis, whereby the fate of a statement lies in the hands of others (TD 106), is now found at the level of action. Hence, the fate of an action lies in the hands of others—human as well as nonhuman actors. For example, with regard to the sociotechnical graph for the hotel key fob Latour states that "the diagram keeps track of successive changes undergone by customers, keys, hotels, and hotel managers" (TD 108). First, this illustrates how executing programs of action is hampered or facilitated by different measures, whether these be material, like the weight of a hotel key fob, or symbolic, like the loss of social prestige. Second, the graphs depict the consecutive changes that the "setting"—the chain of humans and nonhumans (TD 110)—undergoes. This appears as the concrete approach to fulfilling empirically the sociology of technology agenda of *Science in Action*.

However, these are not the only references and connections. In addition to the proximity of this conception to Haraway's *Cyborg Manifesto*, in which the potential of human-machine couplings is illuminated from a Marxist and feminist position, there is again an affinity to Deleuze and Guattari, this time to the theory of (desiring) machines outlined in *Anti-Oedipus.* Far from being an affirmation of a mechanistic view of the unconscious, this theory of machines, which Latour had early taken note of, pursued the goal of developing a new, "transversal" theory of the relationship between the body and technology:

> The object is no longer to compare man and the machine in order to evaluate the correspondences, the extensions, the possible or impossible substitutions of one for the other, but to bring them into communication in order to show how man *is a component part* of the machine, or combines with something else to constitute a machine. The other thing can be a tool, or even an animal, or other men.[10]

Here "machine" does not refer any more to an isolated technical object but denotes a precarious assemblage of heterogeneous partial objects (Latour's "actors") that despite or rather because of internal frictions interact productively.

Have We Never Been Postmodern?

Against this background *We Have Never Been Modern* proposes a new "constitution" for modernity that, in addition to the human actors, gives due consideration to the nonhuman actors. In addition, the book contains a forthright polemic against the discourse of postmodernism. Because of their lack of understanding of the problematic structure of modernity, their "incomplete skepticism" (NB 9), their "despair" and their "nihilism" (NB 134), Latour rejects the theoreticians of postmodernism. In his view they are ultimately part of the problem:

> Postmodernism is a symptom, not a fresh solution. It lives under the modern Constitution, but it no longer believes in the guarantees the Constitution offers. It senses that something has gone awry in the modern critique, but it is not able to do anything but prolong that critique, though without believing in its foundations. Instead of moving on to empirical studies of the networks that give meaning to the work of purification it denounces, postmodernism rejects all empirical work as illusory and deceptively scientistic. (NB 46)

In this passage Latour cites Baudrillard and also Lyotard as examples of postmodern "despair" and in particular Lyotard's *The Postmodern Condition*, which actually has many points of contact with his own works, such as the view that science is an agonistic field or in identifying rhetoric as an important component of scientific practice.

The section on "Semiotic Turns" leaves a similarly ambivalent impression. Here Latour rails against the "philosophies of language, discourse, or texts" (NB 85) that in the past he had made productive use of in his interrogation of science's world of texts and images. Besides Algirdas Julien Greimas, he now accuses Roland Barthes and Jacques Derrida (whom he had cited affirmatively in *Laboratory Life*) of aiding and abetting a variety of linguistic thinking that centers on an "autonomized discourse" (NB 63) that can scarcely be connected to its reference points, its referents, anymore.

From Canguilhem and Bachelard Latour had distanced himself earlier. His reassertion of this in *We Have Never Been Modern* seems all the more strange because Bachelard's pointer that "purification" is of seminal significance in the practice of modern science provided a basic motif of Latour's theory of modernity.[11]

Thus it is surprising to find that in spite of this polemic the tone of Latour's sociology of technology studies is definitely postmodern. Under the pseudonym "Jim Johnson" he indulges in an explicit masquerade and at the same time celebrates in a tongue-in-cheek fashion the disappearance of the author.[12] And the essays in *La clef de Berlin* are replete with puns and allusions.

It starts with the cover of this book, which shows a painting by Jean Lagarrigue of the double-bladed key with the words *Ceci est une clef*, "This is a key." Clearly this is a reference to Foucault. Foucault had analyzed a famous painting by René Magritte that at the top has a realistic depiction of a pipe but underneath the words *Ceci n'est pas une pipe*, "This is not a pipe."[13]

The message of Latour's image asserts that the philosophically notable aspect does not lie in the interaction between text and image (as Foucault had seen it) but in the things themselves; one would be tempted to say in their "materiality," were it not for the pragmaticist concept of a thing that is developed in the book.

Further allusions and irony are aimed at another thinker of postmodernism. The last text in *La clef de Berlin* begins with an interpretation of *The Ambassadors* by Hans Holbein the Younger, a painting that Jacques Lacan had used in his famous seminar, *The Four Fundamental Concepts of Psychoanalysis.*[14] The diagrams and schemata that Latour includes increasingly to illustrate his texts—his "Macintosh doodles"[15]—can also be interpreted as witty responses to the world of proliferating Lacanian graphs.[16] In this connection the apparently arbitrary choice of the door as a sociology of technology example is particularly striking. This is a further case in which Latour singles out an object that had played a prominent role in Lacan's seminars.

"A door must either be open or shut," said Lacan, quoting Littré, in his seminal lecture on "Psychoanalysis and Cybernetics." In his view the door is "a real symbol, the symbol par excellence."[17] As a door has only two possible states, open or closed, it exemplifies the binary code, upon which both the symbolic order (S1/S2) and computers (on/off) are based. Latour makes reference to the same notion: "In computer jargon: A door is exclusively an OR, never an AND" (CB 63). Yet at the same time he appears to contradict this statement by including a comic strip illustration of a door that is at once open *and* closed: the door has a cat flap. Here, too, he appears to be countering the structuralists' thematization of language by bringing in a concrete example from the sphere of technology.

The connection with cybernetics is not random for Latour. Already in *Laboratory Life* he had drawn on concepts from information theory and biophysics to describe the practice of science as a process of transition from "disorder" to "order." And notwithstanding their "irreductionist" orientation, his sociology of technology studies also have a remarkably cybernetic touch. The symmetrical investigation of humans and nonhumans from the perspective of performance could almost be seen as a contribution to the scientific discipline conceived by Norbert Wiener, which sees as its key objective the placing of biological and technological systems on the same level with regard to phenomena of communication and control. Latour comes very close to the operationalism of cybernetics when in this context he introduces a further modifier of what an actor is: "We define an actor or an actant only by its actions in conformity with the etymology" (TD 121). But as in *Laboratory Life* this is not the last word that Latour will have on the science of "Command and Control." In this case, too, he will let literature have last word.

Technology—A Mode of Existence

In fact, Latour's next book is a novel. Published in French in 1992, *Aramis or the Love of Technology* tells the story of a railway transportation system. The name of the system is an acronym of "*A*gencement en *R*ames *A*utomatisées de *M*odules *I*ndépendants dans les *S*tations" (ARAMIS), which can be translated as "arrangement of independent modules into automated trains in stations" (AT 40).

The basic idea of this innovative technology is the nonstop direct transit of passengers to their destinations in customized trains with compartments that can be ordered like a taxi. At the end of the 1960s such a project was begun for the south of Paris. Over the following twenty years the technology was developed and tested and considerable sums invested. Several test tracks with prototypes were built, but the results were not convincing. In 1987 the decision was taken to abandon Aramis, and work on the project ceased.

Shortly afterward RATP, the Paris public transportation company that had been involved in the project, commissioned Latour to do a sociological study of its "failure." From December 1987 to November 1988 he and his assistant Nathaniel Herzberg conducted many interviews with the engineers, managers, public servants, and politicians who had been involved in Aramis and studied the technical reports and documentation.

That in *Aramis* Latour engages with the question of what the conditions are that lead to the success or failure of technical innovations makes it a standard theme for the research conducted at the Center for the Sociology of Innovation at the Écoles de Mines. Beginning with his in-depth study of the potential and difficulties associated with the introduction of electric cars, Callon, the director of the center at the time, had focused on the problems of technical innovations.[18] Latour's *Aramis* book, however, is not standard by any stretch of the imagination. After the textbook-like *Science in Action* and his excursion into the philosophical essay *We Have Never Been Modern*, his new monograph is a mixture of a scientific text and fiction: "scientifiction" (AT ix). In fact *Aramis* is a many-voiced documentary and montage novel that content-wise evokes *Frankenstein* but is also oriented on the classic detective fiction of, for example, Sir Arthur Conan Doyle.

Aramis is described as the "creature" of a group of engineers, managers, and politicians who are attempting to bring this new kind of transportation system to life. Although for a time it looks as though they will be successful, in the end Aramis "dies"—also because this creature, Latour suggests in his "reinterpretation" (PN 280) of Mary Shelley's novel, did not find the love it had hoped for. This is the sense in which the French original's dedication should be understood: "To all those who *loved* Aramis."[19] This suggests that engineers and managers are not only rational people but also potentially crazy visionaries. Further, there are two characters who set out to discover the reasons why Aramis "died." In a duo modeled on Sherlock Holmes and Dr. Watson, the first-person narrator, who introduces himself as an engineering student at the École des Mines, assists a professor of sociology there to solve the case. The professor is referred to as "Norbert H." (AT 300), which can be read as an amalgam of Norbert Wiener and Martin Heidegger.

Unlike the case studies in *La clef de Berlin*, in *Aramis* it is not individual technical objects that are up for discussion but an entire network of railway tracks and circulating compartments, and unlike the myriad extant door openers and speed humps Aramis was quite literally never realized, never became a *thing* in everyday reality. Thus the focus is not the everyday interactions with this nonhuman actor but the negotiations behind the scenes, between engineers, managers, civil servants, and politicians; in other words, the network of oral negotiations, written reports and statements, built test tracks and planned trials. Given this double network Latour's study engages first and foremost with the task of comprehending the specific mode of existence of this particular technical system. What Latour seeks to understand is the special way in which Aramis is "alive," for it did exist for almost twenty years in the twilight zone of blueprints and models.

To accomplish this task Latour does not rely either on Wiener's cybernetics or Heidegger's philosophy. Rather, he uses resources that again bring him into contact with the conceptual and intellectual world of Deleuze and Guattari. He repeatedly refers to Samuel Butler's *Book of the Machines*, which had already served in *Anti-Oedipus* as a reference for overcoming the "classic scheme" that postulates the opposition of mechanism and vitalism.[20] And in particular Latour follows Gilbert Simondon's philosophy of technology, which Deleuze and Guattari often invoke in the context of combining the theory of technical objects and their genealogies with the "prodigious idea of *Nonorganic Life*."[21]

It is more than likely that Latour was also impressed by the fact that Simondon's biologistic treatment of technology is aimed at capturing its specific mode of existence. For although Simondon uses biological terms like "lineage" or "hypertely," it is his explicit intention *not* to equate technological beings with biological beings. On the contrary, he accuses cybernetics of making precisely this mistake. Instead, Simondon intends to characterize technical objects by making critical comparisons, that is, to identify their specificity by establishing phenomenological correspondences *and* differences.[22] The same purpose is served by the differentiating analogies that Simondon draws with other types of objects, for example, religious or aesthetic objects.

Although Latour is very inspired by this schema the approach he actually takes is different. Whereas Simondon engages with the "inner conflicts" of selected technical objects, which he collects, takes apart, and arranges in developmental series (the combustion engine, the vacuum tube, the telephone), the author of *Aramis* concentrates on describing for the railway project the characteristic transitions from signs to things. Instead of interrogating the material culture in a similar way to Simondon, Latour again gives priority to a semiological analysis.[23]

In addition, questions of belief play a role here, too. In connection with the failure of Aramis Latour argues that the engineers in charge were under the spell of the "epistemological myth of a technology" that "can be wholly independent of the rest of society."[24] According to him, they implicitly followed a "Protestant narrative," a model of diffusion. Instead of handing over their idea to other interested parties in the sense of a translation and thereby allowing the idea to be modified, they thus condemned Aramis to death at an early stage.

The alternative would have been the model of a "Catholic narrative," explains Latour. Within the framework of a *récit d'incarnation*, a narrative of incarnation (AT 119), Aramis would have stood a chance of becoming a reality. In other words, the specific mode of existence of a technical object is referred back to a special form of exegesis. For Latour it is the praxis of interpretation that actually decides on existence or nonexistence.[25]

The Agonistic Field Strikes Back

Since his earliest publications Latour had described science as an "agonistic field" in which to speak is to fight and every interpretation (*Ausdeutung*) is exploitation (*Ausbeutung*). In the 1990s this description begins to come true in a rather unexpected way, namely, with reference to Latour's own work. During this decade polemical commentaries about his publications increase appreciably. Up to 1990 his works enjoyed a lively reception in which positive and negative opinions were roughly equal; afterward that year there were frequent cases of commentators brusquely distancing themselves. This reached a high point with the so-called Sokal affair.

In 1996 the American physicist Alan Sokal submitted an article to *Social Text*, an academic journal for cultural studies, ostensibly on the philosophical implications of quantum physics. The day it was published Sokal revealed in another academic journal that his essay was a hoax, a pastiche of postmodernist philosophy larded with scientific nonsense about mathematics and physics quoted from the writings of postmodernist academics. What gave the essay an even greater edge was that it appeared in a special issue of *Social Text* devoted to the so-called science wars, a series of polemical intellectual exchanges carried on from the mid-1990s in academe and the mainstream press on the relationship between left-wing politics, postmodernist philosophy, and science. Sokal's essay cannot be construed as a serious contribution to this debate; its effect was to fuel the polemics.[26]

Latour was one of the authors whom Sokal quoted with feigned approval in his *Social Text* essay. A few years previously Latour had published a sociosemiological analysis of a popular book by Albert Einstein on the special and general theory of relativity. As he had done before in connection with bacteriology and bioinformatics, here, too, Latour attempted to take the idea of science from science—in this case Einstein's possible contribution to understanding epistemological relativism. But unlike the biologist Jonas

Salk, who in the late 1970s had viewed maneuvers of this kind as a promising rapprochement of the social and natural sciences, the physicist Sokal took exception to this procedure and "a lack of precision" in Latour's writing style as well as criticizing that he had misunderstood essential elements of relativity theory.[27]

The debate that ensued, which led to a flurry of publications particularly in the United States and in France, resulted in the discussion about Latour being taken out of the relatively restricted circles of science and technology studies and carried into the public domain. Ironically, the upshot was that Latour was seen as a postmodernist theorist. In *Fashionable Nonsense*, the follow-up book publication to his parody that Sokal wrote together with his friend the physicist and philosopher Jean Bricmont, Latour appears side by side with Lyotard and Baudrillard,[28] precisely the two authors he had distanced himself from in *We Have Never Been Modern*.

In France few were bothered about the affair. Jacques Derrida, who had also been lampooned by Sokal and Bricmont, deplored the general tenor of the debate. Others declared their support, among them Jean-Jacques Salomon, Latour's ex-boss at the Conservatoire National des Arts et Métiers.[29]

At the same time criticism came from a prominent sociologist. David Bloor, the main initiator of the "Strong Program" of the sociology of knowledge and father of the "symmetry principle," published in the middle of the heated debate on Sokal's parody an "Anti-Latour." In a long essay Bloor settles up with Latour for his—in Bloor's view—erroneous account of the Edinburgh School's approach in *We Have Never Been Modern* and elsewhere. Further, Bloor details what he considers are the blatant weaknesses of Latour's approach. A pivotal issue is the attribution of agency to things. In contrast to what Latour says, the Strong Program does recognize agency in nonhuman actors, for example, in the case of entities in physics that resist reliable recording. That Latour goes so far as to attribute agency to door openers or transportation systems is more a metaphysical move than a plausible sociological approach. "It is obscurantism raised to the level of a general methodological principle," bristles Bloor, a point of criticism that is made in substance also by Harry Collins, Steven Yearley, and Timothy Lenoir.[30]

Latour's reaction to his critics always follows the same pattern. Initially, most of the time he admits they are right. Joining forces with Sokal, he rejects "postmodern twaddle" and criticizes social science and humanities journals like *Social Text* which did not at the time operate under the peer-review system. In his reply (together with Callon) to Collins and Yearley, he concedes that some of the publications in question are not empirically valid studies but rather "ontological manifestos."[31] Elsewhere he even regrets "the ridiculous poverty of the ANT vocabulary" (RC 20) that, ultimately, only consists of the terms "association," "translation," and "obligatory passage point." But thereafter Latour invokes the broad institutional basis of the approach that he, Callon, and others are pursuing, which is evidenced by the sheer quantity of publications as well as the number of Ph.D. students at the Center of the Sociology of Innovation. Finally, then, Latour just reiterates the positions expounded in the critiqued publications.

Latour's next book, *Pandora's Hope*, fits into this pattern. The collection of essays published first in English in 1999 can be seen as a considerably expanded and revised version of *La clef de Berlin*, where the majority of sociology of technology contributions have been replaced by history of science studies. It is Latour's final comment on the Sokal debate. Here, too, Latour is struggling to calm things down and to integrate the objections of his critics. For example, to Sokal he addresses the comment that it should be borne in mind that it is thanks to science studies that it has been possible "to interest scores of literary folk in science and technology" (PH 3). In addition, Latour returns to citing Bloor and Collins, he comes back to Bourdieu, and even Roger Guillemin enjoys a surprising comeback on the final pages of the book as his "mentor" (PH 300) in the world of science.

The Crisis of the Networks

Yet it is not only the agonistic field of science that strikes back at Latour's work at this time; in a certain way technology does, too. Latour had repeatedly invoked and described its vitality. At the end of the 1990s technology proved in rather unexpected ways just how lively it was. It became clear to

Latour that rapidly increasing access to the World Wide Web was beginning to threaten the sociological productivity of actor-network theory. In his view the openness and accessibility of the Web invalidated in the long term the notion of the network as developed together with Callon and others:

> At the time, the word network, like Deleuze's and Guattari's term rhizome, clearly meant a series of transformations—translations, transductions—which could not be captured by any of the traditional terms of social theory. With the new popularization of the word network, it now means transport without transformation, an instantaneous, unmediated access to every piece of information. That is exactly the opposite of what we meant. (RC 15)

In addition, Latour's notion of networks came under fire from the work of the sociologists Luc Boltanski and Ève Chiapello. In their study titled *The New Spirit of Capitalism*, published in French in 1999, the two former students of Bourdieu's analyze the "origin of works on networks" and the "naturalization of networks in the social sciences."

According to Boltanski and Chiapello, since the 1960s research on networks has mainly been driven by the sociology of markets and business. In this way the term has accompanied and justified the form of capitalism that Boltanski and Chiapello refer to as the "projective city." By this term the sociologists understand the ongoing social disintegration that is taking place under the guise of enhanced flexibility, which individuals experience as increasingly burdensome and stressful.

Boltanski and Chiapello's main focus is the management texts of business culture from 1960 to 1990, but they also include sociological literature on the subject, including some publications by Latour and Callon. Although they do not expressly criticize it for this, they situate actor-network theory in the vicinity of discourses that contributed to the delegitimation of traditional capitalism with its bureaucracies, nation-states, bourgeois family structures, and social classes. In this context as well the sociological productivity of the network concept appears rather precarious.[32]

However, this is not really the end. Notwithstanding his self-critical reflections, less than two years pass before Latour presents *An Introduction to Actor-Network Theory*, whose main title is *Reassembling the Social*. At the same time Latour's work exudes a spirit of new departure. From the relatively restricted field of the sociology of science and technology he now ventures forth into art, law, and religion. The overarching objective is to profile the regimes of expression, modes of existence, and exegetical forms that go against the grain of what Latour will call henceforth the communication culture of the "double-click."

EIGHT

The Coming Parliament

The last project that the American architect Louis Kahn worked on was a parliament building complex. It had been in planning since the early 1960s. While he was building the institute in La Jolla, California, for Jonas Salk, where a few years later Latour would conduct his ethnological studies for *Laboratory Life*, Kahn was already working on the plans for the National Assembly in Dhaka, the capital of today's Bangladesh. Together with his student Muzharul Islam the architect advanced this mammoth project until his unexpected death from a heart attack in 1974. The Jatiyo Sangsad Bhaban, the National Assembly Building, is the largest national capital complex in the world and besides the Parliament, its centerpiece, it includes hostels, dining halls, and a hospital.[1]

The main building material used for the Salk Institute was reinforced concrete. Kahn used it in the construction of Bangladesh's Parliament but also made extensive use of a very traditional building material, not only on the Indian subcontinent—bricks. Significant sections of the Jatiyo Sangsad complex, like the hostels, are of brickwork. In a famous lecture given at the Pratt Institute in Brooklyn, New York, the year before he died, Kahn reflected on the use of this particular material in a manner that, for today's readers, is evocative of Latour's "active and distributed materialism" (RS 129).

Kahn's focus is the problem of design as a balance between natural forms and human order. Within this context the brick is attributed with its own mode of existence. It appears to be a nonhuman actor. In Latour's *Aramis* the railway transportation system raises its voice from time to time and comments on the progress from its point of view (AT 81–82, 294–296), and with Kahn it is the brick that speaks.

To his audience Kahn declared:

> You cannot design anything without nature helping you. . . . When you want to give something presence you have to consult nature. . . . If you think of brick, for instance, and you consult the orders, you consider the nature of brick. This is a natural thing. You say to brick, "What do you want, brick?" And brick says to you, "I like an arch." And you say to brick, "Look, I want one too, but arches are expensive, and I can use a concrete lintel over you, over an opening." And then you say, "What do think of that, brick?" Brick says, "I like an arch."[2]

We encounter a similar conception of design in Simondon's work, to which, as we have seen, Latour repeatedly refers. For Simondon the brick is first and foremost the most prominent example for elaborating his critique of "hylemorphism." Challenging the Aristotelian view that nature is made of substances composed of form and matter and that the making of bricks demonstrates in a paradigmatic way how a substance (*hylé*) is given a form (*morphè*), Simondon points out the dual nature of this relationship. The form (for example, the mold to make the brick), for a start, consists of matter *and* form. Similarly, substance is not pure substance; even when it is extremely malleable it exhibits a certain form.

The consequence for the design process is that an architect, or an "inventor" of a technical object, must situate themselves beyond *morphè* and *hylé* in a region where form is materialized and at the same time substance is rendered formable. Unlike a sculptor who forms marble according to his own particular vision of a model, and unlike a worker whose product exists in the mind before it is brought forth into reality, for Simondon the creator of a technical object functions as a kind of inductive moderator for whom nature is by no means some uniform substance that can be dominated at will; it is a heterogeneous, often resistant materiality that conveys a multiplicity of "implicit forms."[3]

Consequently, architects and engineers who are striving to arrive at a new object must cooperate with the substance-forms *and* the form-materials. They face the task of entering this prefigured region and moving along its pathways in order to create from out of them.[4] Drawing on Souriau (and Deleuze), Latour describes this process not as construction but as the "*instauration*" of a work of art (or technology).[5] As a result the figure of the designer loses its modernistic connotation. Latour even goes so far as to say that "the more we think of ourselves as designers, the less we think of ourselves as modernizers."[6]

Latour's works from the previous ten-year period can be regarded as a search for the "implicit forms" in the space that knowledge designs and organizes. This is in evidence when he engages with art, religion, and law as well as in his texts on urban planning and architecture.[7] The theme common to all these more recent studies is arguing against the myth and the technique of the "double-click." By this Latour means the idea suggested particularly by the Internet that there really is information that functions purely as messages:

> Double-click communication, this immediate and costless access, this conveyance that appears to demand no transformation, has itself become, for our contemporaries, the model of all possible communication, the ideal, the metric standard of all movement, the judge of all faithfulness, the guarantee of all truth. (TRS 22)

Opposing this conception of communication, which has become dominant, Latour seeks to assert different "forms of speaking" or "regimes of utterance" and corresponding "regimes of truth production."[8] Thereby the materiality of the respective forms of communication, for instance in the legal sphere or religion, is as much of interest as the specific relationship to their subject that these forms imply. In other words, two of Latour's central themes—circulation and reference—are retained in these studies.

Assembling

Like Kahn and his architecture, in Latourian philosophy the path also leads from the laboratory to the parliament. Since the beginning of his collaboration with Callon the idea of a parliament had functioned as a kind of leitmotif in Latour's texts—from the conception of scientific discourses as a shift (or modality) of ways of looking at a problem, to the laboratory scientist as "spokesperson" who tries to convince a skeptical dissenter of the factuality of scientific facts, and to the Serres-influenced understanding of things as public assemblies that have their place in a new, "nonmodern" constitution of modernity. In 1999 Latour devoted an entire book to this leitmotif. *Politiques de la nature*, translated into English as *Politics of Nature*, proposes in the name of "political ecology" to put an end to the modern separation of public life "into two houses"—society or the subject on the one side and nature or the object on the other (PN 13).

This proposition appears like a generalized anthropology of science that is stretching out into politics. But Latour also returns to earlier reflections on the development and transmission of knowledge in the book, which is at the same time a contemporary dialogue with the "cosmopolitan" philosophy of Isabelle Stengers, to whom the book is dedicated.[9] Arguing in favor of a culture of social experimentation, Latour's objective is indeed to redefine the relationship of science and politics:

> we need not political science but science policy, that is, a function that makes it possible to *characterize the relative fruitfulness* of collective experiments, without its being monopolized right away by either scientists or politicians. (PN 202)

This programmatic objective heralds an institutional reorientation that continues to influence Latour's work up to the present day. After nearly twenty-five years at the Centre de Sociologie de l'Innovation, in the autumn of 2006 he became a professor at the leading French-speaking institution for political science, the Institut d'Études Politiques in Paris ("Sciences Po" for short) where from 2007 to 2012 he was also vice president for research. His main field is the development of innovative forms of politics.

Thus it is hardly surprising that in recent years we have seen Latour increasingly in the role of parliamentarian. In 2005 he and Peter Weibel devised the major exhibition *Making Things Public*, which literally gathered together aesthetic, technical, and scientific objects in a parliament of things and also assembled assemblies, from "Thing" sites to religious councils and the parliament buildings of Western democracies.[10] The year before the exhibition Latour was in Paris in his role as president of the Society for Social Studies of Science (4S) and had moderated a plenary session on the subject of "Public Proofs," which was held appropriately in the building of the Senate, the upper house of the French Parliament.

Parallel to these activities Latour is an increasingly "parliamentarian" sociologist and philosopher. In his undiminished prolific output of texts he exhibits a new openness and allows authors whose voices had not been heard in his writings for some time (e.g., Bourdieu and Foucault) to have their say. Looking back to the past, Latour has also begun to situate himself within a history of ideas and theories that was bordered on the one side by the sociology of Gabriel Tarde and the philosophy of Étienne Souriau and on the other by the pragmatism of William James and Alfred N. Whitehead. Looking to the future, Latour's sights are set on a "school of political arts," which will bring together the relatively separate systems of political representation, scientific representation, and aesthetic representation in an interdisciplinary and international endeavor.[11]

Rejoicing

The parliament of things that Latour has been building for some years now is multistoried and spacious. When one bears in mind that the word "parliament" is Middle English from the Old French *parlement*, meaning "speaking," from the verb *parler*, then its cohesion seems assured. The ground for Latour's new orientation on politics was in fact prepared by revisiting facets of earlier concerns.

The first two books he brought out in the first decade of the new millennium, *Rejoicing* and *The Making of Law*, hark back to ethnology and exegesis. The first publication, a sociological report on "the torments of religious speech," as well as the second, an ethnographic study of the Conseil d'État, the French supreme court in administrative law, follow the tried and tested schema of a visit to a specific place, which is then the starting point for an examination of the forms of exegesis practiced there.

Observation of the Conseil d'État begins in the Palais Royal in Paris, where it resides. A body of the French national government, the Council of State acts as legal adviser of the executive branch and as the supreme court for administrative justice. Composed mainly of top-level legal officers, its vice president is the highest-ranking civil servant in France. Based on ethnographic observations made in the Palais Royal in 1996 and 1997 Latour analyzes the kind of transmitting of statements that is practiced there (ML 229).

Rejoicing—the French title reads *Jubiler*—begins in a similar way, with a visit to an almost thousand-year-old church in Montcombroux to study the distinctive quality of religious speech in the age of the Internet but also to present and perhaps even to reenact it. One of the key points made is that religious speech has no reference in the usual sense of the word. It does not correspond to any subject that somehow lies outside this particular mode of speaking or that is at a certain distance to it and can be accessed like a website or document. According to Latour the subject of religious speech lies within itself, in the act of enunciation; it is tied to the performance of speaking. Thus religious speech is "a form of speech that evokes beings who appear and disappear depending entirely on *how they are said*" (TRS 120–121).

Péguy is one of the very few names mentioned in this context. As if Latour wants to revisit his early active engagement with *Clio* and the issue of repetition and resurrection, he states: "There is no religious speech that isn't hesitant, stuttering, embarrassed" (TRS 83). This is actually not only an impartial sociological observation; it is a description of the reflexive style in which Latour wrote this book.

One might even say that the specific layout of *Rejoicing* mimics the form of biblical manuscripts. In the original French edition the text is not divided into chapters, and there are no subtitles or footnotes, no title page, and not even a proper cover. The end of the book confirms this proximity to his earlier concerns in connection with *Clio*. For Latour it is both about rejoicing (*jubiler*) and a jubilee (*jubilé*)—the two thousandth anniversary of the birth of Christ.

Judging

The Making of Law also reconnects with the past, but in a different way and at different location than *Rejoicing*. As Latour explains in the preface, he has arrived at this sociological study of law through his analysis of biological research practices conducted at the Salk Institute: "The book you are about to read is the *Laboratory Life*, not for the construction of facts, but for the construction of legal arguments (*moyens de droit*)" (ML ix). The empirical basis of the study is his privileged access to the presentations and discussions of the petitions submitted to the Conseil d'État. Latour traces the specificity of legal arguments and interpretations whereby Greimas's semiotics, which in the interim had been dismissed by Latour, comes into its own again. In 1976 Greimas had already analyzed legal discourse in his book *Sémiotique et sciences sociales*.[12]

In addition, Latour comes back to his project to write an "ethnography of the filing folder" that he had first drafted in the 1980s, and he devotes an entire chapter to tracking a file on its passage through the rooms of the Conseil in the Palais Royal—from its receipt stamp to preparing it for legal use, the addition of documentation, its labeling and formatting, preparation for the oral proceedings, possible publication, and finally the archiving of the entire thing (ML 70–106). Here, too, one is reminded of an early Resnais documentary short, *Toute la mémoire du monde*, in which the filmmaker shows the long journey of a book from the post room of the National Library via the cataloguing department to its allocated position on a shelf in the stacks.

As in *Rejoicing* the implicit background to Latour's ethnography of the Conseil d'État is the communication regime of the double-click. Thus he writes, for example: "Law, like religion, like politics, deceives those who want to transport information" (ML 268). The reason for this deception, according to Latour, lies in the internal resistance of law to forced forms of technologizing and rationalizing. The "laborious embodiment of meaning," which in legal discourse is achieved through special forms of speech and writing, can therefore not be substituted by calculations of the type that mechanical devices or computers can execute (ML 272).

The special role that this assigns once more to human actors can be seen in a table where Latour compares the "enunciation regimes of law and science" (ML 235). Although the basic difference lies in the chains of "references" (science) and "obligations" (law), the specific differences mentioned include "implicit" versus "explicit work of writing," "inscription by nonhumans" versus "inscription by humans," and "iteration and development of the corpus of knowledge" versus "homeostasis of the corpus of law" (ML 235). This raises the question as to how far differentiation of specific forms of exegesis can in fact be supported by a broad symmetrization of all basic concepts, or whether a kind of preclassification into humans on the one hand and nonhumans on the other is required.

Be that as it may, the comparative table of the two enunciation regimes in *The Making of Law* is the first clear elaboration of a project that Latour has pursued until very recently. Initially presented in 2007 at the colloquium in Cerisy-la-Salle devoted to Latour's work, the goal of this project is the comparative study of "various types of truth production"[13] in science, technology, religion, and law—and also in art and politics. Based on the respective empirical studies, Latour distinguishes between these types of truth production on the basis of translations and transformations as well as their relation to their particular subject.

The *grand tableau* of these types of truth production was published in 2013 under the title *An Inquiry into Modes of Existence*. However, already *The Making of Law* allows for characterizing the general tenor of the corresponding project. Compared to the domination of the double-click, the respective enunciation regimes of law and science comprise specific forms of deceleration. The profiling of this difference is like a warning against the false hope that communication can ever be direct and free from misunderstandings. This is the political and so to speak critical dimension of Latour's ethnological study of the Conseil d'État.

At this point the study converges with Latour's politicoecological vision of a "parliament of things," for there, too, the issue is deceleration: "In order to force ourselves to 'slow down,' we will have to deal *simultaneously* with the 'sciences, with natures, and with politics,' in the plural" (PN 3). Similar to Stengers, Latour insists on the necessity of diplomacy and negotiations with unfamiliar partners.

Walking

Latour's interest in a comparative theory of modes of speaking and corresponding regimes of truth production can be understood as a return to the combination of ethnology and exegesis. At the same time, it prepares the answer to a question that is often put to empiricists: What is the overarching connection between the various studies published so far? What is the "big picture" (PVE 59; RS 187)?

Latour would not be Latour if he did not take this question literally as well. Since the mid-1980s the anthropology of science, which had been significantly shaped by Latour, had claimed images and maps as important, if not crucial, objects of investigation. Latour reconsiders this issue from a new angle when he thematizes the production of "big pictures" with regard to an actor-network model that can be understood as a gigantic summation of alternative models to the networks of the double-click. What is meant is the city of Paris.

Paris ville invisible, a book produced in collaboration with the photographer Emilie Hermant and the graphic designer Susanna Shannon, explores the possibility of devising an appropriate image of this metropolis that exemplifies the culture of modernity up to and including the present day. The title of the large-format volume, copiously illustrated in color, does not so much reference Italo Calvino's *Invisible Cities*, as Latour intimates on the back cover, but instead Alain Touraine's sociological work *La société invisible* (1977).[14] For in *Paris ville invisible* Latour pursues the general question as to the perceptibility of society. To this end the social space of the city is mapped in a "succession of photographic essays" (RS 1).

In sophisticated combinations of text and images the essays represent different "sequences," which at first glance look like suggested city walks for tourists: from the administrative department of the École des Mines to the offices of the French meteorological service and on to a neurobiology laboratory and then to the Grande Galerie of the Natural History Museum, or from Café de Flore via a monitoring station of the waterworks to the central market for fresh foodstuffs.

However, each sequence calls up different representations of the city—street signs, city maps, postcards—and involves different technical infrastructure—Métro lines, water mains, electric power cables. Here urban life is described in terms of its networks and centers of calculation, as Latour described the world of technoscience in *Science in Action*. For "centers of calculation" he introduces the neologism *oligopticon*, which, in contradistinction to Foucault's concept of the panopticon, underlines that what is visible is dependent on local associations and practices (PVE 28; RS 181). Whereas Foucault understands "panopticism"—the power-wielding gaze of surveillance where the operator remains invisible—as a general, characteristic feature of the "disciplinary society,"[15] with the oligopticon Latour is trying to locate what is seemingly global in concrete terms. His goal is a new type of sociology: a situation-oriented "science of living together" (RS 259) that is above all interested in "how to keep the social flat" (RS 165–246); in other words not falling back immediately on the ready-made categories of the established disciplines of history and social science for its concrete descriptions and observations.

The explicit model for proceeding in this way in *Paris ville invisible* is once again a science. In earlier works Latour had referenced bacteriology and bioinformatics positively to set off his own sociological approach; in his book on Paris it is the physiological method of chronophotography that he uses programmatically. In practical terms what Marey called a "photographic gun" is the model that inspired the notion of creating "sequences" (PVE 21). By means of photographs taken in rapid succession Marey had produced studies of "the natural sequence of the movements" of an organism on the plate (PVE 21), for example, a pigeon in flight. In a similar vein, Latour seeks to portray the city's spaces in dynamic succession, interwoven with the changing movements of its inhabitants and visitors and not merely in static juxtaposition.

Incidentally, the idea of using Marey's graphic method in the social sciences was first mooted in the nineteenth century. It was actually considered by Gabriel Tarde, whom Latour claims as an alternative "precursor" for his own sociology (RS 14). Around 1900 Tarde recognized in the graphic technologies as propagated by Marey, Abbé Rousselot, and others that it might be utilized to make sociology "a truly experimental science."[16]

What *Paris ville invisible* creates through its graphic sequencing is understood by Latour as an antipanorama. The total view is rejected, for like the nineteenth-century panoramas it has no visible lacunae and suggests to observers that they are fully immersed in the real world "without any artificial mediations or costly flows of information leading from or to the outside" (RS 188). In fact the praxis of sociology is situated on the same level as all other oligoptics; that is, it works just like a Parisian laboratory or museum with views and perspectives tied to the locality.

This is the point where Latour's study of Paris converges with the introductory exposition in *Reassembling the Social*. In both cases the social cannot be grasped through an isolated image, no matter how large it is. Yet neither can the social be captured independently of every image; it emerges "in the montage of images" and the transformations this effects. It results from "a cross-cutting view, a progression, a formatting, a networking" (PVE 29) of the most diverse actors in their respective environments.

Liquefying

By rejecting the panorama as a possible form of the "big picture" of social science Latour reaffirms his reservations about networks. As we saw above, because of the growing access to the Internet and its burgeoning content Latour became ever more skeptical about the theoretical usefulness of the network concept. He confirms this by increasingly turning to organic images, for example, "rhizome," instead of continuing to espouse the technical understanding of network-like structures.[17]

Ultimately, his reservations about the panoramas designed by social science theories also go in this direction, for there is a close connection between the panoramic view and the technical infrastructure of the railway network. As Dolf Sternberger once remarked, the loss of the optical depth dimension accompanied the panorama, and this was universalized by trains: "The railroad elaborated the new world of experience, the countries and oceans, into a panorama."[18]

Thus the oligoptics also claim to open up a new approach to the theme of the network. Unlike the panorama, they bring out the spaces between the threads, lines, and paths of the network. By creating a montage of single views, which do *not* make up a whole, the oligoptics not only sketch the networks as such but also render comprehensible "their lack of control on what is left in between their networks" (RS 188). Latour calls the substance found there "the plasma" (PVE 4) that designates the sphere "which is not yet formatted, not yet measured, not yet socialized, not yet engaged in metrological chains, and not yet covered, surveyed, mobilized, or subjectified" (RS 244).

The image of the network, which Latour in company with Sloterdijk now qualifies as "anaemic" en passant,[19] has thus become considerably expanded. It encompasses what precedes the networks and what they are based upon. Parallel to this symmetrization the concept of society is also expanded substantially: "Take a map of London and imagine that the social world visited so far occupies no more room than the subway. The plasma would be the rest of London, all its buildings, inhabitants, climates, plants, cats, palaces, horseguards" (RS 244).

Insofar as Latour describes the plasma as fluid and formless,[20] we can detect here a further reverberation of Deleuze and Guattari's philosophy. From the beginning in *Anti-Oedipus* the hard sphere of machines pieced together from partial objects is placed alongside the liquid world of the "body without organs." In Latour's plasma one recognizes this body "with no shape or form" whose distinguishing features are that it is a flow of "amorphous, undifferentiated fluid."[21] There are also allusions to Simondon in as much as this philosopher of technology does not predicate experience of the human body solely on individual organs (ears, eyes, etc.) but includes all the *matière vivante* that surrounds them: "blood, lymph, fascia."[22]

Last, Latour is revisiting a motif of Charles Péguy's. In *Clio* Péguy had illustrated the difference between history and tradition by comparing the modern discipline of history with a long railway line that runs along the coast: traveling on it one can stop at any station one desires. In this metaphor tradition is the coastline with its marshes, its areas where water and land meet.

Against this background it seems insufficient to focus sociology on actors and their networks. For Latour this is now only half of the truth, the truth of modernity. The networks themselves float in plasma, a fluid body, whose dimensions, state, and particularity have yet to be determined. Unlike with a brick, in this case one does not know at all of what nature the relation of *hylé* to *morphè* is.

Summarizing

In his latest book, Latour's parliament building takes on concrete form. *An Inquiry into Modes of Existence* is the revised and expanded version of the manuscript that he presented and discussed at the Cérisy colloquium in 2007. It is, in many respects, a truly remarkable book. First, because of its sheer size. With a total of almost five hundred pages it is the longest text that Latour has ever published. And footnotes, bibliography, glossary, and index are not included. These are only available, after personal registration, via a website that accompanies the book.[23]

With this double form of publishing, the *Inquiry* aims—second—to contribute to current debates on the "digital humanities." Based on the discussion with his readers and on additions to the contents Latour appears to envision publishing a second revised and expanded version of the book. One is tempted to term this process *crowd reviewing.*

Third, the theoretical ambition of the book is quite remarkable. What Latour presents us with is nothing less than the philosophical *summa* of his work up to 2012. It is true that Latour is a philosopher by training, and, as we have seen, from the very beginning his empirical work in sociology and/or anthropology was repeatedly interrupted by theoretical contributions—from the "Irreductions" section of *The Pasteurization of France* to *We Have Never Been Modern.* However, never before has Latour stepped out so clearly as a metaphysical and ontological thinker. The *Inquiry* marks his latest turn to a comprehensive system of categories, a philosophical encyclopedia.

Yet the emphasis on philosophy can also be seen as a somewhat maneuver. By "coming out as a philosopher,"[24] Latour distances himself from the context of actor-network theory and, conversely, intensifies his dialogue with Isabelle Stengers. In fact, ANT continued to be a joint project of Latour, Callon, and others, and in the end its scope remained restricted to sociology. The *Inquiry* transgresses this personal and disciplinary context.

Latour does insist, however, that the big advantage of network analysis is that it "makes it possible to understand through what series of small *discontinuities* it is appropriate to *pass* in order to obtain a certain *continuity* of action" (MoE 33), for example, in a laboratory or in an engineering office. But now the disadvantage of this method is seen as making it almost impossible "to *determine* the type of value that seems to circulate in a particular network and to give it its specific tonality" (MoE 36).

It is Stengers' philosophy, inspired by Deleuze and Guattari as well as by Whitehead, that outlines a new and promising perspective in this respect. It not only opens our eyes to problems of ecology; at the same time it proposes a new form of politics. Following Stengers, Latour adopts one of Whitehead's basic principles: "above all, not to shock common sense" (MoE 59; see also MoE 477). The ultimate goal of Latour's philosophical project is to find "all the ways of *speaking well* about something to someone" (MoE 125). Or, more precisely, in "speaking well in the agora to someone about something that concerns him" (MoE 139). This is the parliamentarian or, at least, the diplomatic perspective of the *Inquiry*, which seeks to capture the specific values tied to specific modes of existence.

Initially, the somewhat bedazzling expression "modes of existence" is only meant to function as a synonym for what Latour had previously termed "regimes of enunciation."[25] In other words, here he introduces another concept that is not clearly defined but employed pragmatically. Its meaning is derived by referring to the empirical studies that he published previously on the different forms of exegesis in science (*Laboratory Life*), technology (*Aramis*), and law (*The Making of Law*).

In a second step, Latour attributes a certain task or function to the expression. By speaking of "modes of existence," he intends to do away with the (modern) separation of language and being, of *les mots et les choses* (MoE 20–21). The parallel he offers is the performative linguistics of John L. Austin who, in *How to Do Things with Words*, transgressed the boundaries between language and action.[26]

The consequences are far reaching: What Latour, ever since his joint work with Fabbri, had conceived of as the "modalizing" of utterances (e.g., the use of indirect speech or graduated judgments) is now rephrased in terms of modalizing a substance. This substance, however, is not continuous or homogeneous but refers to an even more fundamental level (with Deleuze and Guattari one could call it "the plane of immanence")[27] that gives rise to all substances and their respective modes. To this notion Latour attaches an impressive plea against the simple "modern" idea of *matter* as something exterior and dead and argues in favor of transforming this idea into a plural understanding of *materials* (MoE 209).

In order to characterize further the term "modes of existence" Latour refers to Simondon and Souriau.[28] He is certainly correct when he says that Simondon's book on the mode of existence of technical objects has not been read seriously or developed further by the sociology of technology.[29] It is also apposite to point out that Souriau is a largely forgotten writer both inside and outside of France (except, perhaps, for visual studies scholars who might recall his contributions to *filmologie*).[30] However, it seems excessive when Latour highlights the singularity and "idiosyncrasies" of Souriau and underscores his "polar" distance to phenomenology.[31] In fact, both Simondon and Souriau were quite close to phenomenology.

In the 1950s Simondon worked closely with Mikel Dufrenne, who, in his *Phénoménologie de l'expérience esthétique* (Phenomenology of aesthetic experience) deals extensively with the specific ontological status of the aesthetic object.[32] And Souriau, in his truly remarkable book on the different modes of existence (recently republished in French by Latour and Stengers), does not only refer to Husserl and Heidegger in passing. Souriau in particular quotes a student of Husserl *and* Heidegger, Oskar Becker, when he comes to discuss the mode of existence of scientific and other abstract entities, for example, geometrical theorems.[33]

In this 1927 treatise on *Mathematische Existenz* [Mathematical existence], Becker relies on Heidegger's understanding of ontology as a "hermeneutics of facticity"[34] in order to demonstrate that even mathematics can be understood as a way of interpreting *Dasein* in highly specific ways. Roman Ingarden, another student of Husserl, adopted a similar approach when analyzing, also in the late 1920s, the "mode of existence" (*Seinsweise*) of literary works of art.[35] Latour's *Inquiry*, by again stressing the connection between exegesis and ontology, connects with these phenomenological and aesthetic traditions of philosophy—without really being aware of it, it seems.

The starting point for his *Inquiry* is also a rather traditional motif. It figures prominently in Heidegger's "Letter on 'Humanism'" as well as more recently in Peter Sloterdijk's writings.[36] In the introduction, Latour describes the "generalized housing crisis" (MoE 23) of the moderns, who vacillate "between economy and ecology" (MoE 22) without ever really establishing an *oikos*, a household. With rhetorical questions he then legitimizes his project—in spite of all previous criticism of "big pictures"—to draw up an extensive tableau of modes of existence that could function as a new "home" for the moderns:

> Why would we have not the right to propose them [the moderns] a form of habitation that is more comfortable and convenient and that takes into account both their past and their future . . . ? Why would they wander in the permanent utopia that has for so long made them beings without hearth or home—and has driven them for that very reason to inflict fire and bloodshed on the planet? (MoE 22)

Taking into account the table on the final pages of his book (MoE 488–489), however, the architectural structure designed by Latour does not look like a parliament. Rather, it resembles a residential building.

The structure has three main floors:[37] on the first floor we find science (=*Reference*), *Technology*, and *Fiction* (or literature)—three modes of existence that are very familiar from Latour's sociological studies. On the second floor, we encounter some more old friends, *Law*, *Religion*, and *Politics*. And on the top floor reside three new modes of existence. In spite of the criticism that Latour had inflicted upon conventional sociology in the name of ANT, here we find modes of existence that appear to be "purely" social: *Organization*, *Morality*, and *Attachment*.

Besides these three floors, the residential building has a cellar and an attic. The cellar serves as a kind of storage and facility room. It contains materials and immaterials that constitute, as it were, the environment, or *Umwelt*, of the other modes of existence: renewable primary products and animals (*Reproduction*) and extremely lofty entities such as gods and ghosts

(*Metamorphosis*). The material and immaterial substances are spatially organized but also structured in terms of time by objective *Habits*.

Up to the attic. There, we encounter modes of existence that aim to analyze the modes of existence. These are *Networks* but also interpretive keys that allow determination of the tonalities and colors (*Preposition*) of the mode of existence under study. According to Latour, these values become especially clear in conflicts and/or crossovers between two or more modes of existence. The regime of the *Double-Click* also attempts to analyze modes of existence. However, imbued as it is with a "horror of hiatuses" (MoE 487) and convinced that there actually is communication without transformation, it is doomed to fail.

So much for the vertical structure of the house; now to its horizontal structure. On the various floors of the building each mode of existence is studied more closely with respect to a number of key criteria. We can picture this as an allocation of rooms that serve different purposes. To begin with, each mode of existence gets a workroom where it can display the functioning of its networks in terms of continuity (*Trajectory*) and discontinuity (*Hiatus*). Second, a living room is assigned in which the mode of existence can display the *Beings to Institute* or rather the beings it constructs in terms of "instauration." In the case of technology these beings are inventions; in the case of fiction, works of art, etc. Third, a kind of dining room is allocated in order to determine which *Conditions of Felicity* (*and Infelicity*) have to be met so that a mode of existence can fulfill its purpose. And finally, a bathroom or vestibule is provided that pictures the *Alterations* that each mode of existence entails: Does it lead to multiplying worlds (as it is the case in Fiction) or does it ensure the continuity of actions and actors (Law)? Does it calculate the impossible optimum of behavior (Morality)?

Latour's presentation of this structure in the main text of the *Inquiry* does not follow this schematic model, however. It is guided by a different motif, which is also reminiscent of Heidegger but even more of Péguy: "the hiking metaphor" (MoE 74). On numerous occasions Latour describes hiking tours, promenades, and walks, that is, the act of leaving the house in order to explore new places and spaces. His concrete example is a fictitious young anthropologist engaged in studying the moderns. Again and again, she goes out onto the streets (for example, to visit a laboratory) or takes an imaginary walk to the *agora*, which is not just a place for public gatherings but—as Latour points out with etymological expertise—also designates the political sphere in which Categories are formed: "*kata-gorein* is first of all 'how to talk about or against something or someone in public' " (MoE 59).

In other words, Latour's young anthropologist is not a romantic roamer but an authentic *flâneur* in a big city. She often has a map or a city guide in her hands. She carries with her pens and paper—and, most importantly, a questionnaire that structures her anthropological research. This is the concrete meaning of *Inquiry*. The original French expression, *Enquête*, stands for "investigation" and literally for "poll" or "survey." In fact, Latour emphasizes that from the start his own empirical work on the different regimes of truth production was guided by a questionnaire that subjected each regime to the same criteria—from Trajectory to Instauration and Alteration (MoE 100). And there is more.

With the *Inquiry*, Latour also aims at carrying out a survey among his readers. This is another purpose of the interactive website that provides access to the book, including its footnotes, index, and bibliography. Latour invites his readers to post their comments and criticisms, and, in particular, he wants them to answer the question whether or not the descriptions of the modes of existence offered in the book correspond to their own experience. Directly addressing the reader, he announces that "You and other participants . . . will extend the work begun here with new documents, new sources, new testimonials" (MoE xx). The broader perspective is to create, within a digital agora, an author network that actively contributes to preparing an enhanced version of the *Inquiry*.

There is some irony in the fact that of all the possible tools of research, Latour clearly prefers the questionnaire. Since the days of Max Weber the questionnaire is one of *the* classical research instruments of modern sociology, and, as historians have pointed out, the emergence and evolution of this instrument is closely tied to the history of modernity.[38] How can the modern instrument of the questionnaire be in a position to direct empirical studies concerning the anthropology of the moderns? Why would Latour give up the ethnographical principle that in order to investigate a tribe, we should not utilize the concepts, categories, and methods of the tribe itself?

Considering the massive objections that Latour raises to the truth regime of the double-click, one could also ask why it should be the technology of the Internet that allows him to develop his new, pluralist ontology. Or is it simply wrong to assume that the hugely widespread double-click technology fosters the existence of communication without transformation? Then, the paradoxical conclusion would be that in fact the Internet is the coming parliament.

Conclusion

Bruno Latour aims to approach the material featured in his studies with radical openness and curiosity. We don't know, he says, what a society is, what it consists of, and what holds it together.

> We do not have to decide for ourselves what makes up our world, who are the agents "really" acting in it, or what is the quality of the proofs they impose upon one another. Nor do we have to know in advance what is important and what is negligible and what causes shifts in the battle we observe around us. (PF 9)

And we know just as little, he continues, about what surrounds our world, what an environment is, how it works, and what it can do (PN 80, 197). We don't even know what constitutes knowledge. Or at least this is what we have to assume if we want to find out about anything at all (SIA 7).

These are the provocations of Latour's philosophy. They challenge us to leave behind all our everyday knowledge and all our previous experience and dive into a new sociology and a new society. The question as to possible connections with existing theories and areas of practice, with resources of knowledge and accumulated experience, is secondary. What counts is the promise of freshness and freedom, of new observations and new moves.

To keep this promise Latour first constructs a distance that he will later transform into closeness. In his early study on Charles Péguy he used tables and diagrams to create a considerable gap between commentary and what was being commented on to meet the criterion of any "correct exegesis"—not to reiterate what has already been said. The result is an ingenious reading of Péguy that can be understood as a differentiating repetition.

In *Laboratory Life* photographs and diagrams are utilized to achieve the requisite distance to the literary work in the laboratory. At the same time, the authors take great care that their terminology does not falsify the process of how scientific facts are generated. The result is to understand the end products of laboratory work as artistic creations and not as mere fabrications.

In *The Pasteurization of France* and *Aramis* Latour assumes the perspective of a novelist for the purpose of describing the scientific and technical endeavors in question in an appropriate manner, as texts *and* contexts. The result is a new kind of sociology, inspired partly by bacteriology and partly by exegesis, which demonstrates impressively that facts and machines have a life of their own.

Latour is forever taking on different roles—the anthropologist, sociologist, historian, literary writer—so that he can approach the texts and images that fascinate him in the most authentic way possible. One is tempted to see Latour as a disciple of Bertolt Brecht's, who adopts various forms of discourse (sociology, history, literature, and so on) for the purpose of defamiliarizing, as alienation effects, so that other forms of discourse that are standing on the stage with him, as it were (neuroendocrinology, bacteriology, engineering, etc.), can get a better hearing and receive due recognition.[1]

However, Latour chooses to describe his fundamental stance in a different key. He speaks of a generalized agnosticism (PH 275–276). This agnosticism pertains not only to theological precepts like the existence of God but also to scientific presuppositions. Ultimately, what is at stake is a philosophically motivated rejection of all tendencies toward reductionism: *Nothing can be reduced to anything else, nothing can be deduced from anything else, everything may be allied to everything else*—this is indeed the core of Latourian philosophy (PF 163).

Obviously, we are not dealing here with a new edition of Socratic wisdom or Cartesian doubt. Latour is not interested in affirming yet again that we know that we don't know, or in radically questioning reason's potential only to come back to it almost at once with all the more conviction. However, what is also at issue here are not just the transparent methods of the empiricist, who before each new study makes a tabula rasa to set out the previous propositions, substantiate the choice of method, and state the various steps of the procedure employed.

The vital openness and curiosity that Latour exhibits in his case studies and essays should rather be understood as a particular form of philosophical diplomacy. The basic mission of this diplomacy is a noncomprehension that is as intelligent as it is lively. With regard to established theories and findings Latour murmurs to us like Melville's Bartleby: "I would prefer not to."[2]

This is without doubt one of the reasons for Latour's success. It is clearly not sufficient to look for the reasons of this success in pedagogic qualities or in a style that is attractive and very accessible, although both quite obviously play a role. The position that Latour takes up in each new study is neither that of an education professional nor a popularizer. It is in fact quite similar to the position described by Isabelle Stengers, following Deleuze, as that of the "idiot."[3]

Stengers says that the idiot is neither stupid nor a malcontent nor crazy. In the history of philosophy the idiot is frequently someone who doesn't understand Greek. Thus the "conceptual persona" of the idiot is the one who takes a stand against the consensual way that a situation and its inherent logic is presented. Humorously and with idiosyncratic idioms the idiot compels the advocates of the consensus to stop short and start from the beginning again: "The idiot demands that we slow down, that we don't consider ourselves authorized to believe we possess the meaning of what we know."[4]

Thus Latour's openness and freshness, his "unforseeable naïveté" and "disarming benevolence," as Sloterdijk puts it,[5] should be understood as a productive braking maneuver. Latour wants to take us forward by stopping us from carrying on as before. He lets us mark time so that we can then leap.

The respite that this provides contains a seemingly simple gesture. It consists in pointing a finger at something: a building, a painting, a map, a line, a word. This movement of the hand is found in almost all of Latour's books and many of his essays. A laboratory scientist points at a glowing screen displaying a curve that needs explaining. A king places his finger on an outdated map before sending his geographer on a scientific voyage to explore the country. An angel sits by the side of an empty grave, pointing simultaneously to the emptiness below and the richness above.

The meaning of this repeated gesture is obvious. It stands for "*Ego, hic, nunc*" (PVE 17). In other words, it indicates a person, a place, or a point in time. Although in this sense it is self-referential, for Latour it is the social gesture par excellence. The pointing finger not only refers to an "I" but equally to a "You"; it does not merely point toward a specific space and a present but also in a direction; it points to a past and a future.

At the same time the gesture wavers between force and meaning. In fact in classical antiquity the pointing finger signified a transmission of power; the connotation of meaning was added later. And that is only the top layer of this minimal sociosemiotic structure. The basic framework of the pointing gesture is an assemblage of the living and the nonliving, of the body and things, of the human and the nonhuman. Perhaps that is why it seems so self-evident to delegate the function of a pointing finger to things, like pointers, arrows, or signboards.

For Latour this is profoundly fascinating. Just as scientists and technologists continually point to their evidence (curves, sketches, plans), in his studies he refers time and again to the act of pointing. This can be seen as the silent counterpart to the exhortation "Follow the actors!" and also the starting point for developing networks ("Point to the pointers!").

And that is why the things of pointing—Heidegger's "*Zeigzeug*"[6] or signaling equipment—in Latourian philosophy are elementary fixtures of sociality. The social for Latour possesses an eminently referential character.

This demonstrates Latour to be a willing and able pupil of Charles Péguy. In his extensive "Note on Bergsonian Philosophy" Péguy devotes a long passage to signposts, which culminates in praising pointing, ritual, and the community:

> The Catholic is a fellow who walks along a country road and is happy about the signpost which stands there for everyone. And not only that: he does not consult these signposts, which are there for everyone, to find his way. For he knows his way. . . . He consults the signposts to feel a certain kind of joy, a sort of ritual joy of the country road, to practice a certain rite, a rite of the way.[7]

Latour conveys this kind of ritual joy when on his walks through Paris he describes traffic signs, street signs, and commemorative plaques as the basis upon which to develop an "oligoptical" sociology. The effect is similar when he publicizes a photograph that he took of a signpost encountered in the Engadin Valley in Switzerland not far from Nietzsche's beloved village Sils-Maria. The individual signs of the bilingual signpost point right and left, front and back, up and down, evoking exactly the kind of steroscopic inspection that Latour has frequently emphasized is a characteristic of his "symmetrical anthropology."[8]

Now there are forms of tradition that assure they will be handed down because they do not employ word-for-word repetitions but instead new versions and variants of what has already been said. Inspired by Péguy, it is perfectly clear to Latour that paeans of praise to what is new are not necessarily novel. Breaking with tradition also has a tradition.

This is the sense in which this book has shown how Latour's studies of science and technology—in spite of an ongoing dialogue with the philosophy of Deleuze and Guattari—are focused on *the problem of exegesis*. By virtue of this focus Latour indirectly connects to a twentieth-century context in German-speaking countries, which at a relatively early point in time raised objections to the totalitarianism of scientific claims to objectivity.

During their years together in Marburg, Rudolf Bultmann and Martin Heidegger were convinced that these claims had been accepted in a largely uncritical way by the Marburg school of neo-Kantianism. Instead of this absolutization of science Bultmann and Heidegger offered religion, and also art and technology, as autonomous forms of existence that, in their view, were at the same time autonomous forms of interpreting and forging meaning.

Latour's "anthropology of science" fits into this context, although it is not oriented on the existential analysis of the 1920s but rather on the ethnomethodology of the period around 1970, and although it does not chart the geophilosophical terrain of the German-speaking countries but moves along the Californian coast, from San Diego to Stanford and back again.

Essentially, this "anthropology" is about doing away with the opposition between science on the one side and religion, art, and technology on the other, by understanding science itself as a specific form of exegesis. It follows that there is not only one single form of objectivity—scientific objectivity—but a whole series of objectivities or regimes of truth production, which are based on certain types of exegesis and because of their specificity cannot be reduced one to the other.

This is also evidenced by Latour's approach, his particular way of accessing themes and problems. It is true that he goes back to Nietzsche in order to combine the problem of exegesis with the issue of relating and differentiating "forces." And via Nietzsche (and Lyotard) he establishes a connection to the Sophists, who pursued the objective of creating a symmetry out of opposing viewpoints and thus aimed less at truth and more at effect.

However, Latour's interest in the technical and religious aspects of science is different, more directly motivated. One could almost speak of a theologicophenomenological impetus. Thus, he says

> Science deserves better than naive worship and naive contempt. Its regime of visibility is as uplifting as that of religion and art. The subtlety of its traces requires a new form of care and attention. It requires—why abstain from the word?—yes, spirituality.[9]

As a consequence, it is hardly surprising that the things of science and of technology do not appear in Latour's works predominantly in their materiality. Obviously, an object like Guillemin and Schally's thyrotropin-releasing factor can be manipulated and transported, and obviously, a railway transportation system like Aramis can literally be realized; that is, it can be turned into a thing, a material entity. But this is not Latour's primary concern.

His most important consideration is the endeavor to grasp the special mode of existence of scientific and technological things: as threshold entities betwixt and between the fabricated and the factual, the historical and the logical, the profane and the divine. His anthropology is focused on events that occur in a certain form of exegesis, on differences in inscriptions, which in some measure change both the text and the reader as well as the library where reading and interpretation takes place.

Of course Latour is not the first and not the only scholar who has focused on the specificities of scientific and technological things. In the wake of Husserl and Heidegger, philosophers have analyzed the mode of existence or *Seinsweise* of mathematical entities and works of art.[10] Historians of science have described scientific objects as " 'superreal,' non-natural" things and drawn attention to the "incompleteness" of their effects and actual nature. And philosophers of science have discussed these objects as "the creation of phenomena" in scientific practice.[11]

Latour is one of the very few authors who speaks of transcendence and spirituality in this context. Contemporary history of science has considerable difficulties with this idiom, as does media studies, for example; even when, or particularly when it is the "trace assemblage" of the experiment (Rheinberger) or the "media-technological apriori" of literature (Kittler) that is at issue—seemingly quite close to Latour.[12]

However, the most productive contributions in the area where history of science and media studies overlap focus on examining the material culture and active corporeality of epistemic, technical, and aesthetic practices. The potential of this materialism, which approaches the question of difference and repetition more from the point of view of machine production than from the angle of literary representations and regards scientific facts more in the sense of effects than in the sense of objects has by no means been exhausted. This is evidenced by the exemplary work of Peter Galison and Simon Schaffer as well as by the recent studies by Myles Jackson, Hannah Landecker, and Cornelia Vismann, to name only a few.[13]

Latour's actuality and his problematic lie in different terrain. In a situation where keywords such as "religious wars," "fundamentalism," and "creationism" are very much to the fore and where the ecological crisis increasingly affects our daily life, Latour's stereoscopic view of modernity and tradition, of science and religion, as well as of history and the gospels proves to be an enormously fruitful heuristic for the present day. This was demonstrated in striking ways by *Iconoclash*, the exhibition co-curated by Latour and Peter Weibel on the world "beyond the image wars in science, religion, and the arts," and it is demonstrated again by his recent book and digital platform on the multiple modes of existence.

Confronted by the almost excessive fruitfulness of these projects, however, one does recall Deleuze's reservations about Péguy's philosophy as a culmination of Kantianism. The crucial question is whether it is instructive to speak of *belief* in connection with science. It is also questionable whether one can clearly distinguish different regimes of expression or exegetical ways of being when the basis for such a distinction is tied to the symmetrical examination of *all* actors involved.

Latour provides no answer to these questions. He is too much the masked philosopher, the thinker on a stage. His aim is not a representative sociological survey, a historical reconstruction, or a causal and explanatory account. Ultimately, he says, his concern is not even philosophy but to write books.[14] Latour wants a new richness for descriptions—lively depictions of unfamiliar spaces, discourses, and realities. His concern is to dissuade us away from anthropology as well as theology, away from the discipline of history as well as from the history of science. The glad tidings he brings are that the openness of the event is perennial.

TIMELINE

1947

Bruno Latour is born on June 22 in Beaune, Burgundy (Département Côte d'Or), south of Dijon. His family are wine growers, and their estate Maison Louis Latour has existed since 1797 (not to be confused with the famous Château Latour in Pauillac near Bordeaux).

1966–72

Studies philosophy at the Université de Bourgogne, Dijon, including with the Bultmann specialist André Malet.

1972

Graduates (*agrégation*) in philosophy; grammar school teacher in Gray (Département de la Haute Saône); first studies of Charles Péguy.

1973–75

Development sociology fieldwork for a study initiated by the Office de la Recherche Scientifique et Technique Outre-Mer (ORSTOM) in Abidjan, Republic of Côte d'Ivoire (Ivory Coast). Makes the acquaintance of the ethnologist Marc Augé. At the same time Latour works on his Ph.D. thesis in philosophy.

1975

Ends his studies by gaining a doctorate at the University of Tours. Latour's Ph.D. thesis (*thèse de troisième cycle*) is supervised by the theologian Claude Bruaire. Entitled *Exegèse et ontologie à propos de la resurrection* [Exegesis and ontology with reference to the Resurrection] the dissertation engages with Péguy's *Clio* and the textual structure of the New Testament.

1975–77

Postgraduate research in the laboratory of the French neuroendocrinologist Roger Guillemin (Nobel Prize in Physiology or Medicine 1977, together with Andrew V. Schally and Rosalyn Yalow) at the Salk Institute for Biological Studies in La Jolla, a neighborhood in San Diego, California.

1976

Meets the British sociologist of science Steve Woolgar.

1977–81

Assistant to Jean-Jacques Salomon at the Conservatoire National des Arts et Métiers (CNAM) in Paris. Organizes the newly established study course "Science, Technologie, Société." Collaboration with Michel Callon from the Centre de Sociologie de l'Innovation at the École Nationale Supérieure des Mines. In addition, from 1978 Latour collaborates on the research project organized by Claire Salomon-Bayet of the Centre National de la Recherche (CNRS) on Pasteur and French medicine of the period 1871–1919.

1979

Laboratory Life: The Social Construction of Scientific Facts [co-author Steven Woolgar] (2nd edition 1986, with the title *Laboratory Life: The Construction of Scientific Facts*, French edition 1988).

1979–84

Founds and oversees together with Callon the Bulletin *Pandore* (twenty-five issues) as well as the eponymous book series.

1982–91

Professor at the Centre de Sociologie de l'Innovation of the École des Mines.

1984

Les microbes: Guerre et paix, suivi de Irréductions (English edition 1988, titled *The Pasteurization of France*).

1987

Science in Action: How to Follow Scientists and Engineers Through Society (French edition 1989). Habilitation at the École des Hautes Études en Sciences Sociales.

1991

Nous n'avons jamais été modernes: Essai d'anthropologie symmétrique (English edition 1993).

1992–2006

Tenured professor at the Centre de Sociologie de l'Innovation. Dialogue with Michel Serres, *Eclaircissements* (1992, English edition 1995). *Aramis ou l'amour des techniques* (1993, English edition 1996).

1993

La clef de Berlin et autres leçons d'un amateur de sciences.

1996

Beginning of the Sokal affair.

1998

Paris ville invisible (co-author Emilie Hermant).

1999

Pandora's Hope: Essays on the Reality of Science Studies (French edition 2001) and *Politiques de la nature: Comment faire entrer les sciences en démocratie* (English edition 2004).

2002

Jubiler ou les tourments de la parole religieuse (English edition 2013) and *La fabrique du droit: Une ethnographie du Conseil d'État* (English edition 2010). Exhibition *Iconoclash*, at the Center for Art and Media (ZKM), Karlsruhe (curated in collaboration with Peter Weibel, Peter Galison, Dario Gamboni, Joseph Leo Koerner, Adam Lowe, and Hans-Ulrich Obrist).

2003–5

President of the Society for Social Studies of Science (4S).

2005

Reassembling the Social: An Introduction to Actor-Network Theory (French edition 2007). Exhibition *Making Things Public: Atmospheres of Democracy* at the Center for Art and Media (ZKM), Karlsruhe (curated in collaboration with Peter Weibel and many others).

2006

Professor at the Institut d'Études Politiques de Paris (Sciences Po). *Chroniques d'un amateur des sciences.*

2007

Director of Research and Vice President of Sciences Po. Cerisy Colloquium "Exercises de métaphysique empirique," lectures and discussions on the manuscript "Résumé d'une enquête sur les modes d'existence ou bref éloge de la civilisation qui vient."

2010

Cogitamus: Six lettres sur les humanités scientifiques. "Selon Bruno Latour," a series of events at the Centre Georges Pompidou. Inauguration of the Sciences Po Ecole des Arts Politiques (SPEAP).

2012

Enquête sur les modes d'existence: Une anthropologie des Modernes (English edition 2013).

NOTES

INTRODUCTION

1. The epigraph to this chapter is from Gilles Deleuze and Félix Guattari, *A Thousand Plateaus: Capitalism and Schizophrenia*, trans. Brian Massumi (Minneapolis: University of Minnesota Press, 1987), 372.

2. http://www.bruno-latour.fr/.

3. "Most Cited Authors of Books in the Humanities," *Times Higher Education* (March 26, 2009), http://www.timeshighereducation.co.uk/story.asp?storyCode=405956.

4. Robert Crease, Don Ihde, Casper Bruun Jensen, and Evan Selinger, "Interview with Bruno Latour," in *Chasing Technoscience: Matrix for Reality*, ed. Don Ihde and Evan Selinger (Bloomington: Indiana University Press, 2003), 15.

5. References to frequently cited works by Latour are given in the text by an abbreviation and page numbers; see the List of Abbreviations for Frequently Cited Works on the first pages of this volume. Unless otherwise indicated, all translations are ours (H.S./G.C.).

6. Robert Spaemann, *Der Ursprung der Soziologie aus dem Geist der Restauration. Studien über L.G.A. de Bonald* (Munich: Kösel, 1959).

7. Sigmund Freud, *Jokes and Their Relation to the Unconscious* [1905], in *The Standard Edition of the Complete Psychological Works of Sigmund Freud*, ed. James Strachey, 6th printing (London: Hogarth Press/Institute of Psychoanalysis, 1973), 34; Alexandre Koyré, "Traduttore—Traditore: A propos de Copernic et de Galilée" [1943], in *Etudes d'histoire de la pensée scientifique* (Paris: Gallimard, 1973), 272–274.

8. Marshall McLuhan, *Understanding Media: The Extensions of Man*, 6th ed. (Cambridge, Mass.: The MIT Press, 1997), 90.

9. Jean-François Lyotard, *The Differend: Phrases in Dispute*, trans. Georges Van Den Abbeele (Minneapolis: University of Minnesota Press, 1988), 17.

10. T. Hugh Crawford, "An Interview with Bruno Latour," *Configurations* 1, no. 2 (1993): 262. On Latour and Deleuze see François Dosse, *Gilles Deleuze and Félix Guattari: Intersecting Lives*, trans. Deborah Glassman (New York: Columbia University Press, 2010), 511–515.

11. Deleuze and Guattari, *A Thousand Plateaus*, 7.

12. Ibid., 372, 19.

13. On Tarde, see, for example, Gilles Deleuze, *Difference and Repetition*, trans. Paul Patton (New York: Columbia University Press, 1994), 25–26, 76; Deleuze and Guattari, *A Thousand Plateaus*, 216–219. On Péguy, see Chapter 1 of this volume; on Simondon, see Chapter 7.

14. Bruno Latour, "For David Bloor . . . and Beyond: A Reply to David Bloor's 'Anti-Latour'," *Studies in History and Philosophy of Science* 30, no. 1 (1999): 115.

1. EXEGESIS AND ETHNOLOGY

1. On Marey, see, for example, François Dagognet, *Etienne-Jules Marey: A Passion for the Trace*, trans. Robert Galeta with Jeanine Herman (New York: Zone, 1992).

2. Bruno Latour, Philippe Mauguin, and Geneviève Teil, "A Note on Socio-Technical Graphs," *Social Studies of Science* 22 (1992): 33–57.

3. Peter Sloterdijk, "Ein Philosoph im Exil oder: Der Mann, der die Wissenschaften liebt," tribute speech at the presentation of the Siegfried Unseld Prize to Bruno Latour, Frankfurt am Main, September 28, 2008.

4. Clive Coates, *Côte d'Or: A Celebration of the Great Wines of Burgundy* (London: Weidenfeld and Nicolson, 1997), 179–180.

5. On Roupnel, see Philip Whalen, "'A Merciless Source of Happy Memories': Gaston Roupnel and the Folklore of Burgundian Terroir," *Journal of Folklore Research* 44, no. 1 (2007): 21–40. On the French tradition of philosophy of science, see Jean-François Braunstein, "Bachelard, Canguilhem, Foucault: Le 'style français' en épistémologie," in *Les philosophes et la science*, ed. Pierre Wagner (Paris: Gallimard, 2002), 920–963.

6. See, for example, Geoff Bowker and Bruno Latour, "A Becoming Discipline Short of Discipline: (Social) Studies of Science in France," *Social Studies of Science* 11 (1987): 715–748.

7. See Jean Brun, *Les conquêtes de l'homme et la séparation ontologique* (Paris: Presses Universitaires de France, 1961). On Brun, see Maryvonne Perrot, ed., *Journée Jean Brun: Dijon, le 18 mars 1995* (Dijon: Editions Universitaires de Dijon, 1996), with an introduction by Canguilhem and an essay by Dagognet, "Jean Brun devant la technique" (11–30).

8. André Malet, *Mythos et logos: La pensée de Rudolf Bultmann* (Geneva: Labor et Fides, 1962). Translated into English as *The Thought of Rudolf Bultmann*, trans. Richard Strachan, preface by Rudolf Bultmann (Garden City, N.Y.: Doubleday, 1971).

9. André Malet, *Le traité théologico-politique de Spinoza et la pensée biblique* (Paris: Les Belles Lettres, 1966); Rudolf Bultmann, *Histoire de la tradition synoptique*, trans. André Malet (Paris: Seuil, 1973). The original German text is *Geschichte der synoptischen Tradition* (Göttingen: Vandenhoeck & Ruprecht, 1921), translated into English as *History of the Synoptic Tradition*, trans. John Marsh (Oxford: Blackwell, 1963). On Malet, see *André Malet ou Un homme en quête de Dieu. Hommages de l'Université de Bourgogne*, ed. Marie-Françoise Conrad, Chantal Picard, and Muriel Vautrin (Dijon: Editions Universitaires de Dijon, 1991), which includes contributions by Nicole Malet, Jean Brun, and Paul Ricœur.

10. Bruno Latour, personal communication, July 16, 2008.

11. T. Hugh Crawford, "An Interview with Bruno Latour," *Configurations* 1, no. 2 (1993): 250; Bruno Latour, "Coming out as a Philosopher," *Social Studies of Science* 40, no. 4 (2010): 600; See also Bruno Latour, *Exegèse et ontologie à propos de la resurrection*, thèse de 3ème cycle, supervised by Claude Bruaire (University of Tours, 1975).

12. André Malet, *Bultmann et la mort de Dieu: Présentation, choix de textes, biographie, bibliographie* (Paris: Seghers, 1968), 68.

13. Charles Péguy, *Clio: Dialogue de l'histoire et de l'âme païenne*, in *Œuvres en prose, 1909–1914* (Paris: Gallimard, 1961), 286.

14. Charles Péguy, "De la situation faite à l'histoire et à la sociologie dans les temps modernes," in *Œuvres en prose, 1898–1908* (Paris: Gallimard, 1959), 997.

15. Hella Tiedemann-Bartels, *Verwaltete Tradition: Die Kritik Charles Péguys* (Freiburg: Alber, 1986), 226.

16. Péguy, *Clio*, 210, 213.

17. Charles Péguy, "Un nouveau théologien: M. Fernand Laudet," in *Œuvres en prose, 1909–1914* (Paris: Gallimard, 1961), 951.

18. Bruno Latour, "Trains of Thought: Piaget, Formalism, and the Fifth Dimension," *Common Knowledge* 6, no. 3 (1997): 179.

19. Gilles Deleuze, *Difference and Repetition*, trans. Paul Patton (New York: Columbia University Press, 1994), 1–27.

20. Péguy, *Clio*, 126.

21. Deleuze, *Difference and Repetition*, 1.

22. Péguy, *Clio*, 180.

23. Deleuze, *Difference and Repetition*, 1.

24. Ibid., 5.

25. Ibid., 22.

26. Ibid., 8.

27. Ibid., 22. See also Leo Spitzer, "Zu Charles Péguys Stil," in *Stilstudien: Zweiter Teil, Stilsprachen* (Munich: Hueber, 1928), 301–364.

28. Marc Augé, *Non-Places: Introduction to an Anthropology of Supermodernity*, trans. John Howe (London: Verso, 1995); and Marc Augé, *In the Metro*, trans. Tom Conley (Minneapolis: University of Minnesota Press, 2002). Augé concluded his extensive field studies of the Alladian people in 1974 and published the results in his dissertation *Théorie des pouvoirs et idéologie* (Paris: Hermann, 1975). Latour attended the celebration organized to mark this event, which took place at the shores of the lake where the Alladians had settled. In June 1974 he witnessed the tragic accident there in which Boniface Ethé Neuba was killed, an informant and consultant of Augé's for many years (SIA 203–204).

29. Augé, *Théorie des pouvoirs et idéologie*, 419–420; See also Marc Augé, ed., *La construction du monde: Religion, représentations, idéologie* (Paris: Maspéro, 1974).

30. Marc Augé, *Pouvoirs de vie, pouvoirs de mort: Introduction à une anthropologie de la répression* (Paris: Flammarion, 1977), 13.

31. Gilles Deleuze and Félix Guattari, *Anti-Oedipus: Capitalism and Schizophrenia*, trans. Robert Hurley, Mark Seem, and Helen R. Lane (Minneapolis: University of Minnesota Press, 1992), 51–137.

32. Gilles Deleuze and Félix Guattari, "Balance Sheet–Program for Desiring Machines," trans. Robert Hurley, *Semiotext(e)* 2, no. 3 (1977): 129–130. This article was added as an appendix to the second French edition of *Anti-Oedipus* in 1972.

2. A PHILOSOPHER IN THE LABORATORY

1. Oswald W. Grube, "Die Geburt des modernen Forschungsbaus in den USA," in *Entwurfsatlas Forschungs- und Technologiebau*, ed. Hardo Braun and Dieter Grömling (Basel: Birkhäuser, 2005), 22.

2. See http://www.harpercollins.com/author/microsite/about.aspx?authorid=24395.

3. See, for example, Roman Jakobson, "Linguistics in Relation to Other Sciences" [1967–1970], in *Selected Writings* (The Hague: Mouton, 1971), 2:655–696; Roman Jakobson, "The Fundamental and Specific Characteristics of Human Language" [1969], in *Selected Writings*, ed. Stephen Rudy (Berlin: Mouton, 1985), 7:93–97. See also Lily Kay, *Who Wrote the Book of Life? A History of the Genetic Code* (Stanford, Calif.: Stanford University Press, 2000), 297–307.

4. Edgar Morin, *California Journal*, trans. Deborah Cowell (Portland, Ore.: Sussex Academic Press, 2008).

5. Bruno Latour, "Biographie d'une enquête—à propos d'un livre sur les modes d'existence," *Annales de philosophie* 75, no. 4 (2012): 553.

6. Roger Guillemin, "[Autobiography]," in *The History of Neuroscience in Autobiography*, ed. Larry R. Squire (San Diego: Academic Press, 1998), 2:117.

7. Ibid., 2:118.

8. Arne Naess [Ness], *Erkenntnis und wissenschaftliches Verhalten* (Oslo: Dybwad, 1936), 9; Stewart E. Perry, *The Human Nature of Science: Researchers at Work in Psychiatry* (New York: Free Press, 1966); Georges Devereux, *From Anxiety to Method in the Behavioral Sciences* (The Hague: Mouton, 1967). On Perry, see Ron Westrum, "[Review of *Laboratory Life* and *The Human Nature of Science*]," *Science Communication* 3 (1982): 437–440.

9. Michel de Certeau, "La rupture instauratrice ou le christianisme dans la culture contemporaine," *Esprit* n.s. 6 (1971): 1178.

10. Ibid., 1191, 1197.

11. Ibid., 1197.

12. Georges Thill, *La fête scientifique: D'une praxéologie scientifique à une analyse de la décision chrétienne* (Paris: Aubier-Montaigne, 1973).

13. François Dosse, *Michel de Certeau: Le marcheur blessé* (Paris: La Découverte, 2002), 367.

14. Thill, *La fête scientifique*.

15. Peter Weingart, *Wissenschaftssoziologie* (Bielefeld: Transcript, 2003), 15.

16. Robert K. Merton, *The Sociology of Science: Theoretical and Empirical Investigations* (Chicago: University of Chicago Press, 1973).

17. Guillemin, "[Autobiography]," 113.

18. Roger Guillemin, personal communication, January 8, 2011.

19. Georges Canguilhem, *La formation du concept de réflexe aux XVIIe et XVIII siècles*, 2nd ed. (Paris: Vrin, 1977), 154.

20. Hebbel E. Hoff and Roger Guillemin, "The First Experiments on Blood Transfusion in France," *Journal of the History of Medicine and Allied Sciences* 18 (1963): 103–124; Claude Bernard, *Cahier rouge*, trans. Hebbel E. Hoff, Lucienne Guillemin, and Roger Guillemin (Cambridge, Mass.: Schenkman, 1967); Claude Bernard, *Lectures on the Phenomena of Life Common to Animals and Plants*, vol. 1, trans. Hebbel E. Hoff, Roger Guillemin, and Lucienne Guillemin (Springfield, Ill., 1974).

21. Hebbel E. Hoff, Leslie A. Geddes, and Roger Guillemin, "The Anemograph of Ons-en-Bray: An Early Self-Registering Predecessor of the Kymograph with Translations of the Original Description and a Biography of the Inventor," *Journal of the History of Medicine and Allied Sciences* 12, no. 4 (1957): 424–448.

22. François Cusset, *French Theory: How Foucault, Derrida, Deleuze, and Co. Transformed the Intellectual Life of the United States*, trans. Jeff Fort (Minneapolis: University of Minnesota Press, 2008), 69.

23. Jean-François Lyotard, *The Postmodern Condition: A Report on Knowledge*, trans. Geoff Bennington and Brian Massumi (Minneapolis: University of Minnesota Press, 1993), 51, 45. For Lyotard's 1975 lectures on Nietzsche and the sophists, see Keith Crome, *Lyotard and Greek Thought: Sophistry* (Houndmills: Palgrave Macmillan, 2004).

24. Lyotard, *The Postmodern Condition*, 24 (endnote 87, quotation 92). Some years later this preference will bring them together again, in the exhibition *Les immatériaux* curated by Lyotard in Centre Georges Pompidou. Latour participated in a computer-assisted collaborative writing project along with Michel Butor, Jacques Derrida, Isabelle Stengers, and others, which sought to explore possible collective definitions of philosophical concepts; see Jean-François Lyotard and Thierry Chaput, eds., *Épreuves d'écriture* (Paris: Ed. du Centre Pompidou, 1985).

25. Lyotard, *The Postmodern Condition*, 10.

26. Paolo Fabbri, personal communication, January 13, 2011.

27. On Fabbri, see Yves Jeanneret, "La provocation sémiotique de Paolo Fabbri," *Communication & langages* 146 (2005): 129–141; 148 (2006): 117–135.

28. Pierre Bourdieu, "Le champ scientifique," *Actes de la recherche en sciences sociales* 2, nos. 2–3 (1976): 88–104.

29. Georges Canguilhem, *The Normal and the Pathological*, trans. Carolyn R. Fawcett, in collaboration with Robert S. Cohen (New York: Zone, 1989), 239.

30. Algirdas Julien Greimas, *Sémiotique et sciences sociales* (Paris: Seuil, 1976).

31. Thomas Bernhard, *The Loser*, trans. Jack Dawson (New York: Knopf, 1991).

3. MACHINES OF TRADITION

1. On the PAREX project see Aant Elzinga, "Some Notes from the Past," *EASST Review* (June 1997), http://www.easst.net/review/june1997/elzinga.shtml.

2. Steve W. Woolgar, "The Identification and Definition of Scientific Collectivities," in *Perspectives on the Emergence of Scientific Disciplines*, ed. Gérard Lemaine, Roy MacLeod, Michael Mulkay, and Peter Weingart (The Hague: Mouton, 1976), 234–245.

3. Michael J. Mulkay and David O. Edge, "Cognitive, Technical, and Social Factors in the Growth of Radio Astronomy," in *Perspectives on the Emergence of Scientific Disciplines*, ed. Gérard Lemaine, Roy MacLeod, Michael Mulkay, and Peter Weingart (The Hague: Mouton, 1976), 153–186. On the issue of reflexivity, see John H. Zammito, *A Nice Derangement of Epistemes:*

Postpositivism in the Study of Science from Quine to Latour (Chicago: University of Chicago Press, 2004), 234–239.

4. See G. Nigel Gilbert and Steve Woolgar, "The Quantitative Study of Science: An Examination of the Literature," *Science Studies* 4 (1974): 279–294; Michael J. Mulkay, G. Nigel Gilbert, and Steve Woolgar, "Problem Areas and Research Networks in Science," *Sociology* 9 (1975): 187–203; and the already quoted article by Woolgar, "The Identification and Definition of Scientific Collectivities."

5. Derek J. de Solla Price, "Networks of Scientific Papers," *Science* 149, no. 3683 (1965): 510–515.

6. Gilbert and Woolgar, "The Quantitative Study of Science," 283; Mulkay, Gilbert, and Woolgar, "Problem Areas and Research Networks in Science," 187.

7. Gilbert and Woolgar, "The Quantitative Study of Science," 289.

8. See Karin Knorr-Cetina, *The Manufacture of Knowledge: An Essay on the Constructivist and Contextual Nature of Science* (Oxford: Pergamon, 1981). Knorr-Cetina's book is based on field studies that she carried out from October 1976 to October 1977 at the Western Regional Research Center, a federal research facility for agricultural and food research in Berkeley, California.

9. Arie Rip, "Citation for Bruno Latour, 1992 Bernal Prize Recipient," *Science, Technology, and Human Values* 18, no. 3 (1993): 379.

10. Louis Althusser, "Présentation," published as the first section of the article by Pierre Macherey, "La philosophie de la science de Georges Canguilhem: Epistémologie et histoire des sciences," *La pensée* 113 (1964): 50.

11. In a programmatic essay of 1976 Serres had appropriated the perspective of biophysics. See Michel Serres, "The Origin of Language: Biology, Information Theory, and Thermodynamics," in *Hermes: Literature, Science, Philosophy*, ed. and trans. Josue V. Harari and David F. Bell (Baltimore, Md.: Johns Hopkins University Press 1982), 71–83.

12. Gilles Deleuze and Félix Guattari, *Anti-Oedipus: Capitalism and Schizophrenia*, trans. Robert Hurley, Mark Seem, and Helen R. Lane (Minneapolis: University of Minnesota Press, 1992), 3; Gilles Deleuze and Félix Guattari, "Balance Sheet-Program for Desiring Machines," trans. R Hurley, *Semiotext(e)* 2, no. 3 (1977): 131.

13. Jacques Derrida, *Of Grammatology*, trans. Gayatri C. Spivak (Baltimore, Md.: Johns Hopkins University Press, 1976).

14. François Dagognet, *Écriture et iconographie* (Paris: Vrin, 1973).

15. The famous phrase is: "Instruments are nothing but theories materialized." See Gaston Bachclard, *The New Scientific Spirit* [1934], trans. Arthur Goldhammer (Boston: Beacon, 1984), 13.

16. Georges Canguilhem, *Knowledge of Life* [1952], ed. Paola Marrati and Todd Meyers, trans. Stefanos Geroulanos and Daniela Ginsburg (New York: Fordham University Press, 2008), xx.

17. Besides cybernetics, the mention of black-boxing refers to Richard D. Whitley, "Black Boxism and the Sociology of Science: A Discussion of the Major Developments in the Field," *Sociological Review Monographs* 18 (1972): 61–92; see also LL1 24.

18. Donna Haraway, "[Review of Bruno Latour and Steve Woolgar, *Laboratory Life*]," *Isis* 71, no. 3 (1980): 489.

4. PANDORA AND THE HISTORY OF MODERNITY

1. Jean-François Lyotard, "Considerations on Certain Partition Walls as the Potentially Bachelor Elements of a Few Simple Machines," in *Le machine celibi/The Bachelor Machines* [exhibition catalog], ed. Harald Szeemann (Venice: Alfieri, 1975), 100.

2. Ibid., 102.

3. See, for example, Michel Callon, "Les firmes multinationales: Un théâtre d'ombres," *Sociologie du travail* 2 (1974): 113–140; Lucien Karpik, "Organizations, Institutions, and History," in *Organization and Environment: Theory, Issues, and Reality*, ed. Lucien Karpik (Beverly Hills, Calif.: Sage, 1978), 15–68; Alain Touraine, "The Firm: Power, Institution, and Organization," in *The Post-Industrial Society: Tomorrow's Social History—Classes, Conflicts, and Culture in the Programmed Society*, trans. Leonard F. X. Mayhew (New York: Random House, 1971), 139–192. On Callon, see also Madeleine Akrich et al., eds., *Débordements: Mélanges offerts à Michel Callon* (Paris: Transvalor–Presses des Mines, 2010).

4. Anonymous, "Pandore," *Pandore* 19 (1982): 55.

5. Michel Callon and Bruno Latour, "Introduction," in *La science telle qu'elle se fait: Anthologie de la sociologie des sciences de langue anglaise*, ed. Michel Callon and Bruno Latour (Paris: La Découverte, 1982), vii.

6. David Bloor, *Knowledge and Social Imagery* (London: Routledge & Kegan Paul, 1976); David Bloor, *Socio/logie de la logique: Ou, les limites de l'épistémologie*, trans. Dominique Ebnother (Paris: Pandore, 1983).

7. G. Nigel Gilbert and Michael Mulkay, *Opening Pandora's Box: A Sociological Analysis of Scientists' Discourse* (Cambridge: Cambridge University Press, 1984), 1–2.

8. Bruno Latour, "Avertissement de l'auteur pour l'édition française," in *L'espoir de Pandore: Pour une version réaliste de l'activité scientifique* (Paris: La Découverte, 2007), 5.

9. In this connection, see the suggested "Return to the Sophists" in Isabelle Stengers, *The Invention of Modern Science*, trans. Daniel W. Smith

(Minneapolis: University of Minnesota Press, 2000), 161–166; more generally on this theme, see Barbara Cassin, *L'effet sophistique* (Paris: Gallimard, 1995).

10. See Georges Canguilhem, *Ideology and Rationality in the History of the Life Sciences* [1977], trans. Arthur Goldhammer (Cambridge, Mass.: The MIT Press, 1988).

11. See, for example, Jean-Jacques Salomon, *Science et politique* (Paris: Seuil, 1985); Jean-Jacques Salomon, "Georges Canguilhem ou la modernité," *Revue de métaphysique et de morale* 90, no. 1 (1984): 52–62.

12. Claire Salomon-Bayet, *L'institution de la science et l'expérience du vivant: Méthode et expérience à l'Académie royale des sciences, 1666–1793* (Paris: Flammarion, 1978).

13. Georges Canguilhem, "Bacteriology and the End of Nineteenth-Century 'Medical Theory,'" in *Ideology and Rationality*, 51–77.

14. Ibid., 54. See also François Dagognet, *Méthodes et doctrines dans l'œuvre de Pasteur* (Paris: Presses Universitaires de France, 1967).

15. Claire Salomon-Bayet, "Présentation," in *Pasteur et la révolution pastorienne*, ed. Claire Salomon-Bayet (Paris: Payot, 1986), 12; cf. the review by Helga Nowotny, "The Beginnings of Scientific Modernity," *Social Studies of Science* 17 (1987): 753–759.

16. Bruno Latour, "Pasteur and Pouchet: The Heterogenesis of the History of Science," in *A History of Scientific Thought: Elements of a History of Science*, ed. Michel Serres (Oxford: Blackwell, 1995), 526–555.

17. Translator's note: In the past, "*dispositif*" in the Foucauldian sense was often translated as "apparatus," among other things. However, considerable dissatisfaction with this translation has developed. See, e.g., Giorgio Agamben, *What Is an Apparatus? and Other Essays*, trans. David Kishik and Stefan Pedatella (Stanford, Calif.: Stanford University Press, 2008); Jeffrey Bussolini, "What Is a Dispositif?" *Foucault Studies* 10 (2010): 85–107. The French original is used here to signal its status as a definite, independent concept.

18. Michel Foucault, "The Confession of the Flesh," in *Power/Knowledge: Selected Interviews and Other Writings*, ed. Colin Gordon, trans. Colin Gordon, Leo Marshall, John Mepham, and Kate Soper (Sussex: Harvester, 1980), 194.

19. See, however, Philipp Sarasin, *Reizbare Maschinen: Eine Geschichte des Körpers 1765–1914* (Frankfurt: Suhrkamp, 2001).

20. Gilles Deleuze, "Ethology: Spinoza and Us," trans. Robert Hurley, in *Incorporations*, ed. Jonathan Crary and Sanford Kwinter (New York: Zone, 1992), 625–633.

21. Ibid., 630.

22. Michel Callon, "L'opération de traduction comme relation symbolique," in *Incidence des rapports sociaux sur le développement scientifique et technique: Séminaire de recherche tenu à la Maison des Sciences de l'Homme 1974/1975*, ed. Claude Gruson (Paris: CORDES, 1976), 123.

23. Ibid. Here Callon does not yet refer to Serres but to Jean-Paul Sartre with regard to the philosophy of translation. According to Callon, Sartre's autobiographical work *The Words* demonstrates that the author understands each and every reading to be a translation and a rewriting.

24. See also Bruno Latour, "Insiders and Outsiders in the Sociology of Science; or, How Can We Foster Agnosticism?" in *Knowledge and Society: Studies in the Sociology of Culture Past and Present*, ed. Robert A. Jones and Henrika Kucklick (Greenwich, Conn.: JAI, 1981), 3:200–216.

25. Gilles Deleuze and Félix Guattari, *What Is Philosophy?*, trans. Hugh Tomlinson and Graham Burchell (New York: Columbia University Press, 1994), 207.

5. OF ACTANTS, FORCES, AND THINGS

1. Simon Schaffer, "The Eighteenth Brumaire of Bruno Latour," *Studies in History and Philosophy of Science* 22, no. 1 (1991): 192.

2. Michel Serres, *Statues: Le second livre des fondations* (Paris: Flammarion, 1987), 111. See also NB 83.

3. Bruno Latour, "Postmodern? No, Simply *A*modern! Steps Towards an Anthropology of Science," *Studies in History and Philosophy of Science* 21, no. 1 (1990): 161.

4. Georges Canguilhem, "L'histoire des sciences dans l'œuvre épistémologique de Gaston Bachelard" [1963], in *Etudes d'histoire et de philosophie des sciences* [1968], expanded ed. (Paris: Vrin, 2002), 186.

5. Vincent Descombes, *Modern French Philosophy*, trans. L. Scott-Fox and J. M. Harding (Cambridge: Cambridge University Press, 1980), 85, 91.

6. See Michel Serres, "Commencements," *Le Monde* (January 1, 1980); as well as Ilya Prigogine, Isabelle Stengers, and Serge Pahaut, "La Dynamique: De Leibniz à Lucrèce," *Critique* 380 (January 1979): 35–55.

7. Michel Serres, "Platonic Dialogue," in *Hermes: Literature, Science, Philosophy*, trans. and ed. Josué V. Harari and David F. Bell (Baltimore, Md.: Johns Hopkins University Press, 1982), 66 (the text dates from 1966). For an extensive commentary on this text, see Bruce Clarke, "Noise and Form: Michel Serres's Cybernetics in Autopoietic Context," paper given at the Max Planck Institute for the History of Science, Berlin, Germany, January 18, 2011.

8. Translator's note: In French, "parasite" has the additional meaning of *bruit indésirable*, such as interference or static.

9. Michel Serres, *The Parasite*, trans. L. R. Schehr (Baltimore, Md.: Johns Hopkins University Press, 1982), 226.

10. Ibid.

11. The second sentence is taken almost verbatim from the entry for "Actor" in Algirdas Julien Greimas and Joseph Courtés, eds., *Semiotics and Language: An Analytical Dictionary* (Bloomington: Indiana University Press, 1988), 7.

12. See Bruno Latour and Madeleine Akrich, "A Summary of a Convenient Vocabulary for the Semiotics of Human and Nonhuman Assemblies," in *Shaping Technology, Building Society: Studies in Sociotechnical Change*, ed. Wiebke E. Bijker and John Law (Cambridge, Mass.: The MIT Press, 1992), 259–264.

13. Vladimir J. Propp, *Morphology of the Folktale*, trans. Laurence Scott (Bloomington: Indiana University Press, 1958).

14. Étienne Souriau, *Les deux cent mille situations dramatiques* (Paris: Flammarion, 1950). On Souriau, see Bruno Latour and Isabelle Stengers, "Le Sphinx de l'œuvre," introduction to Étienne Souriau, *Les différents modes d'existence, suivi de l'œuvre à faire* (Paris: Presses Universitaires de France, 2009), 1–75.

15. Algirdas Julien Greimas, *Structural Semantics: An Attempt at a Method*, trans. Daniele McDowell, Ronald Schleifer, and Alan Velie (Lincoln: University of Nebraska Press, 1983), 200–202.

16. That this would instead inevitably lead to misconceptions was quickly recognized; see Hans Ulrich Gumbrecht, "Algirdas Julien Greimas," in *Französische Literaturkritik der Gegenwart in Einzeldarstellungen*, ed. Wolf-Dieter Lange (Stuttgart: Kröner, 1975), 326–350.

17. Karin Knorr-Cetina, "Germ Warfare," *Social Studies of Science* 15 (1985): 581.

18. Gilles Deleuze, *Nietzsche and Philosophy* [1962], trans. Hugh Tomlinson (London: Athlone, 1983), 3.

19. Ibid., 73.

20. David Bloor, "Anti-Latour," *Studies in History and Philosophy of Science* 30, no. 1 (1999): 97.

21. On Nietzsche as an ethnologist of the university, see Friedrich Kittler, *Optical Media: Berlin Lectures 1999*, trans. Anthony Enns (Cambridge: Polity, 2010), 20–21. On Nietzsche's visits to the biological laboratory of the oceanographic observatory at Villefranche-sur-Mer, see Richard Frank Krummel, "Joseph Paneth über seine Begegnung mit Nietzsche in der Zarathustra-Zeit," *Nietzsche-Studien* 17 (1988): 478–495.

22. Graham Harman, *Prince of Networks: Bruno Latour and Metaphysics* (Melbourne: Re.Press, 2009), 12–13.

23. Michel Serres, "Auguste Comte auto-traduit dans l'encyclopédie," in *La Traduction: Hermes III* (Paris: Minuit, 1974), 159.

24. Ibid., 181.

25. Auguste Comte, *Cours de philosophie positive*, ed. Michel Serres, François Dagognet, and Allal Sinaceur, 2nd ed. (Paris: Herman, 1998), 1:534.

26. Jean-François Braunstein, “La philosophie des sciences d'Auguste Comte,” in *Les philosophes et la science*, ed. Pierre Wagner (Paris: Gallimard, 2002), 798.

27. See the essays on Comte in Georges Canguilhem, *Etudes d'histoire et de philosophie des sciences* [1968], expanded ed. (Paris: Vrin, 2002), 61–98.

28. Michel Serres, “Paris 1800,” in *A History of Scientific Thought*, ed. Michel Serres (Oxford: Blackwell, 1995), 448.

29. Michel Serres, “Introduction,” in *A History of Scientific Thought*, ed. Michel Serres (Oxford: Blackwell, 1995), 6.

30. Walter Benjamin, “Eduard Fuchs, Collector and Historian,” in *Selected Writings*, vol. 3: *1935–1938*, trans. E. Jephcott, ed. H. Eiland and M. W. Jennings (Cambridge, Mass.: Belknap Press of Harvard University Press, 2002), 283.

31. Bruno Latour, “Pasteur and Pouchet: The Heterogenesis of the History of Science,” in *A History of Scientific Thought*, ed. Michel Serres (Oxford: Blackwell, 1995), 526–555.

32. Ibid., 553, 554.

33. Ibid., 554.

34. Bruno Latour, “Do Scientific Objects Have a History? Pasteur and Whitehead in a Bath of Lactic Acid,” trans. Lydia Davis, *Common Knowledge* 5, no. 1 (1993): 78.

35. Ibid., 86.

36. Ibid., 88–89.

6. SCIENCE AND ACTION

1. An early example is John Law, “Notes on the Theory of the Actor-Network: Ordering, Strategy, and Heterogeneity,” *Systems Practice* 5 (1992): 379–393. See also especially John Law and John Hassard, eds., *Actor Network Theory and After* (Oxford: Blackwell, 1999); and Latour's *Introduction to Actor-Network-Theory* (=RS) published in 2005.

2. Bruno Latour and Michael Callon, “Unscrewing the Big Leviathan: How Actors Macrostructure Reality, and How Sociologists Help Them to Do So,” in *Advances in Social Theory and Methodology: Toward an Integration of Micro- and Macro-Sociologies*, ed. Karin Knorr-Cetina and Aaron V. Circourel (Boston: Routledge & Kegan Paul, 1981), 277–303.

3. Françoise Bastide died in 1988. Latour dedicated a study to her that was published in 1990, “The Force and the Reason of Experiment” (FR), and he was the editor of a volume of her selected writings in Italian, to which Paolo Fabbri contributed the introduction. See Françoise Bastide, *Una notte con*

Saturno: Scritti semiotici sul discorso scientifico (Rome: Meltemi, 2001), with a bibliography of her works (301–304). See also the commentary by Greg Myers that follows the translation of Bastide's essay, "A Night with Saturn," *Science, Technology, and Human Values* 17, no. 3 (1992): 277–280.

4. Michel Callon, John Law, and Arie Rip, eds., *Mapping the Dynamics of Science and Technology: Sociology of Science in the Real World* (London: Macmillan, 1986). The glossary is on xvi–xvii.

5. Tom Wolfe, *The Right Stuff* (New York: Farrar, Straus and Giroux, 1979); Tracy Kidder, *The Soul of a New Machine* (Boston: Little, Brown, 1981).

6. See, for example, Jean-François Lyotard, *The Postmodern Explained: Correspondence 1982–1985*, trans. Don Barry et al. (Minneapolis: University of Minnesota Press, 1993), 20.

7. Jean-François Lyotard, *The Postmodern Condition: A Report on Knowledge*, trans. Geoff Bennington and Brian Massumi (Minneapolis: University of Minnesota Press, 1993), 46.

8. Thomas S. Kuhn, *The Structure of Scientific Revolutions*, 4th ed. (Chicago: University of Chicago Press, 2012), 112–114.

9. Translator's note: *le physiographe*, invented by the French physicians Delapierre and Vesque in the 1880s, does not translate into English as "physiograph," which is not found either in the *OED* or in Merriam-Webster's. Here "physiograph" has been taken over from the French and anglicized.

10. Jack Goody, *The Domestication of the Savage Mind* (Cambridge: Cambridge University Press, 1977), 3.

11. Thomas P. Hughes, *Networks of Power: Electrification in Western Society, 1880–1930* (Baltimore, Md.: Johns Hopkins University Press, 1993).

12. Bruno Latour, "Drawing Things Together," in *Representation in Scientific Practice*, ed. Michael Lynch and Steve Woolgar (Cambridge, Mass.: The MIT Press, 1990), 19–68.

13. Alexandre Koyré, "Attitude esthétique et pensée scientifique" [1955], in *Etudes d'histoire de la pensée scientifique* (Paris: Gallimard, 1973), 275–288; Edmund Husserl, *L'origine de la géometrie*, trans. Jacques Derrida (Paris: Presses Universitaires de France, 1962).

14. Gilles Deleuze and Félix Guattari, *Anti-Oedipus: Capitalism and Schizophrenia*, trans. Robert Hurley, Mark Seem, and Helen R. Lane (Minneapolis: University of Minnesota Press, 1992), 243.

15. Latour, "Drawing Things Together," 29.

16. Bruno Latour, "[Review of Elizabeth L. Eisenstein, *The Printing Press as an Agent of Change*, Cambridge University Press, 1979]," *Pandore* 13 (1981): 19.

17. Elizabeth L. Eisenstein, *The Printing Press as an Agent of Change: Communications and Cultural Transformations in Early Modern Europe*, vol. 1/2 (Cambridge: Cambridge University Press, 1979), 326, 704.

7. QUESTIONS CONCERNING TECHNOLOGY

1. See Bernward Joerges, ed., *Technik im Alltag* (Frankfurt: Suhrkamp, 1988); and Bernward Joerges and Ingo Braun, eds., *Technik ohne Grenzen* (Frankfurt: Suhrkamp, 1994).

2. Bernward Joerges, personal communication, April 27, 2011.

3. Donna Haraway, "A Cyborg Manifesto: Science, Technology, and Socialist-Feminism in the Late Twentieth Century," in *Simians, Cyborgs, and Women: The Reinvention of Nature* (New York: Routledge, 1991), 152.

4. This publication has not been translated into English. See, however, Bruno Latour, "The Berlin Key, or How to Do Words with Things," in *Matter, Materiality, and Modern Culture*, ed. Paul M. Graves-Brown (London: Routledge, 2000), 10–21.

5. Steven Shapin and Simon Schaffer, *Leviathan and the Air-Pump: Hobbes, Boyle, and the Experimental Life* (Princeton, N.J.: Princeton University Press, 1985).

6. Zygmunt Bauman, *Modernity and Ambivalence* (Cambridge, Mass.: Polity, 1991); Ulrich Beck, *Risk Society: Towards a New Modernity*, trans. Mark Ritter (London: Sage, 1994; originally published in German in 1989); Anthony Giddens, *The Consequences of Modernity* (Cambridge, Mass.: Polity, 1990).

7. Shapin and Schaffer, *Leviathan and the Air-Pump*, 344.

8. Jim Johnson (i.e., Bruno Latour), "Mixing Humans and Nonhumans Together: The Sociology of a Door-Closer," *Social Problems* 35, no. 3 (1988): 298–310.

9. Bruno Latour and Madeleine Akrich, "A Summary of a Convenient Vocabulary for the Semiotics of Human and Nonhuman Assemblies," in *Shaping Technology, Building Society: Studies in Sociotechnical Change*, ed. Wiebke E. Bijker and John Law (Cambridge, Mass.: The MIT Press, 1992), 259.

10. Gilles Deleuze and Félix Guattari, "Balance Sheet-Program for Desiring Machines," trans. R. Hurley, *Semiotext(e)* 2, no. 3 (1977): 117–118.

11. Gaston Bachelard, *Le matérialisme rationnel* (Paris: Presses Universitaires de France, 1953), 78: "Each epoch of science, in its modern development, has established a sort of corpus of substances constituted at a well-defined level of purification."

12. Johnson (i.e., Latour), "Mixing Humans and Nonhumans Together."

13. Michel Foucault, *This Is Not a Pipe*, trans. and ed. by James Harkness (Berkeley: University of California Press, 1983).

14. Jacques Lacan, *The Four Fundamental Concepts of Psychoanalysis: The Seminar Book XI*, ed. Jacques-Alain Miller, trans. Alan Sheridan (New York: Norton, 1981).

15. Bruno Latour, "For David Bloor . . . and Beyond: A Reply to David Bloor's 'Anti-Latour,'" *Studies in History and Philosophy of Science* 30, no. 1 (1999): 115.

16. For example, the scheme concerning the "reduction" and "amplification" of signs in Latour's article about the sampling of soil in the Amazon forest (PH 24–79) resembles the scheme of image and gaze in Lacan, *The Four Fundamental Concepts*, 105.

17. Jacques Lacan, *The Ego in Freud's Theory and in the Technique of Psychoanalysis: The Seminar Book II*, ed. Jacques-Alain Miller, trans. Sylvana Tomaselli (New York: Norton, 1988), 301–302.

18. Michel Callon, *Le véhicule électrique: Un enjeu social*, 4 vols. (Paris: Ecole des Mines, 1978).

19. Bruno Latour, *Aramis, ou l'amour des techniques* (Paris: La Découverte 1992), 7. The English edition is dedicated to Simon Schaffer; see AT VI.

20. See Gilles Deleuze and Félix Guattari, *Anti-Oedipus: Capitalism and Schizophrenia*, trans. Robert Hurley, Mark Seem, and Helen R. Lane (Minneapolis: University of Minnesota Press, 1992), 284–285.

21. Gilles Deleuze and Félix Guattari, *A Thousand Plateaus: Capitalism and Schizophrenia*, trans. Brian Massumi (Minneapolis: University of Minnesota Press, 1987), 411.

22. On this point, see Henning Schmidgen, "Machine Cinematography," *INFLeXions* 5 (2012), http://www.inflexions.org/n5_schmidgenhtml.html.

23. On this point, see also Mark E. Hansen, *Embodying Technesis: Technology Beyond Writing* (Ann Arbor: University of Michigan Press, 2000), 41–47.

24. Bruno Latour, "Ethnografie einer Hochtechnologie: Das Pariser Projekt 'Aramis' eines automatischen U-Bahn-Systems," in *Technografie: Zur Mikrosoziologie der Technik*, ed. Werner Rammert and Cornelius Schubert (Frankfurt: Campus, 2006), 52–53.

25. In spite of such adaptations Simondon's philosophy remains of great significance for Latour's interrogation of technology—and not only of technology. See, for example, Simondon's idea of helping the "technical beings" gain cultural recognition through the commitment of informed representatives. This idea is nothing less than a version of "the parliament of things" limited to technology; see Gilbert Simondon, *Du mode d'existence des objets techniques* [1958], 3rd ed. (Paris: Aubier, 1989), 148.

26. On the "Science Wars" and the Sokal affair see, for example, Keith Parsons, ed., *The Science Wars: Debating Scientific Knowledge and Technology* (Amherst, N.Y.: Prometheus, 2003).

27. Bruno Latour, "A Relativistic Account of Einstein's Relativity," *Social Studies of Science* 18 (1988): 3–44; see Alan Sokal and Jean Bricmont, *Fashionable*

Nonsense: Postmodern Intellectuals' Abuse of Science (New York: Picador, 1998), 92–99. Other physicists were not so dismissive of Latour's reading of Einstein, e.g., N. David Mermin, "What's Wrong with This Reading?" *Physics Today* 50, no. 1 (1997): 92–95.

28. Sokal and Bricmont, *Fashionable Nonsense*, 134–138, 147–153.

29. Jacques Derrida, "Sokal et Bricmont ne sont pas sérieux," *Le Monde* (November 20, 1997); Jean-Jacques Salomon, "L'éclat de rire de Sokal," *Le Monde* (January 31, 1997).

30. See David Bloor, "Anti-Latour," *Studies in History and Philosophy of Science* 30, no. 1 (1999): 97; and the two essays by Harry Collins and Steven Yearley, "Epistemological Chicken" and "Journey Into Space," in *Science as Culture and Practice*, ed. Andrew Pickering (Chicago: University of Chicago Press, 1992), 301–326, 369–389; Timothy Lenoir, "Was the Last Turn the Right Turn? The Semiotic Turn and A. J. Greimas," *Configurations* 2, no. 1 (1994): 119–136.

31. Bruno Latour, "Y a-t-il une science après la Guerre Froide?" *Le Monde* (January 18, 1997); Michel Callon and Bruno Latour, "Don't Throw the Baby out with the Bath School! A Reply to Collins and Yearley," in *Science as Culture and Practice*, ed. Andrew Pickering (Chicago: University of Chicago Press, 1992), 344.

32. Luc Boltanski and Ève Chiapello, *The New Spirit of Capitalism*, trans. Gregory Elliott (London: Verso, 2005), 141–151. Latour and Callon are mentioned on 145.

8. THE COMING PARLIAMENT

1. Mateo Kries, Jochen Eisenrund, and Stanislaus Moos, eds., *Louis Kahn: The Power of Architecture* (Weil am Rhein: Vitra Design Museum, 2012).

2. Louis I. Kahn, "Lecture at the Pratt Institute (1973)," in *Essential Texts*, ed. Robert Twombly (New York: Norton, 2003), 269–271. See also Réjean Legault, "Louis Kahn and the Life of Materials," in *Louis Kahn: The Power of Architecture*, ed. Mateo Kries, Jochen Eisenrund, and Stanislaus Moos (Weil am Rhein: Vitra Design Museum, 2012), 219–234.

3. Gilbert Simondon, *L'individu et sa genèse physico-biologique* [1964] (Grenoble: Millon, 1995), 53.

4. Ibid.

5. Bruno Latour, "Reflections on Etienne Souriau's *Les différrents modes d'existence*," in *The Speculative Turn: Continental Materialism and Realism*, ed. Levi Bryant, Nick Srnicek, and Graham Harman (Melbourne: Re.press, 2011), 309. The concept of instauration plays a major role in MoE 151–178. On Deleuze's reading of Souriau and his use of "instauration," see Leonard Lawler, "A Note on the Relation Between Étienne Souriau's *L'instauration*

philosophique and Deleuze and Guattari's *What Is Philosophy?*" *Deleuze Studies* 5, no. 2 (2011): 400–406.

6. Bruno Latour, "A Cautious Prometheus? A Few Steps Toward a Philosophy of Design (with Special Attention to Peter Sloterdijk)," in *Networks of Design: Proceedings of the 2008 Annual International Conference of the Design History Society (UK), University College Falmouth, 3–6 September 2009*, ed. Fiona Hackne, Jonathn Glynne and Viv Minto (Boca Raton: Universal Publishers, 2009), 3.

7. See particularly PV and PVE as well as Bruno Latour and Albena Yaneva, "'Give Me a Gun and I Will Make All Buildings Move': An ANT's View of Architecture," in *Explorations in Architecture: Teaching, Design, Research*, ed. Reto Geiser (Basel: Birkhäuser, 2008), 80–89.

8. See TRS 120–121 as well as Bruno Latour, "Preface to the English Edition," in *The Making of Law: An Ethnography of the Conseil d'Etat*, trans. Marina Brilman and Alain Potage (Cambridge, Mass.: Polity, 2010), x.

9. Isabelle Stengers, "The Cosmopolitical Proposal," in *Making Things Public: Atmospheres of Democracy*, ed. Bruno Latour and Peter Weibel (Cambridge, Mass.: The MIT Press, 2005), 994–1003.

10. Latour and Weibel, eds., *Making Things Public.*

11. See Bruno Latour and Valérie Pihet, "Sciences Po School of Political Arts," http://www.betonsalon.net/spip.php?article218.

12. Algirdas Julien Greimas, *Sémiotique et sciences sociales* (Paris: Seuil, 1976), 79–128.

13. Bruno Latour, "Coming Out as a Philosopher," *Social Studies of Science* 40, no. 4 (2010): 601.

14. Alain Touraine, *La société invisible: Regards 1974–1976* (Paris: Seuil, 1977).

15. Michel Foucault, *Discipline and Punish: The Birth of the Prison*, trans. Alan Sheridan (New York: Vintage, 1995), 195–228.

16. Gabriel Tarde, *Social Laws: An Outline of Sociology*, trans. Howard C. Warren (London: Macmillan, 1899), 198. Many thanks to Robert M. Brain, University of British Columbia, for drawing my attention to this passage.

17. See, for example, Bruno Latour, "Social Theory and the Study of Computerized Work Sites," in *Information Technology and Changes in Organizational Work*, ed. Wanda J. Orlikowski (London: Chapman & Hall, 1996), 295–307.

18. Dolf Sternberger, *Panorama of the Nineteenth Century*, trans. Joachim Neugroschel (Cambridge Mass.: The MIT Press, 1977), 39. For an in-depth study, see Wolfgang Schivelbusch, *The Railway Journey: The Industrialization and Perception of Time and Space in the Nineteenth Century* (Berkeley: University of California Press, 1986).

19. Bruno Latour, *Changer de société, refaire de la sociologie* (Paris: La Découverte, 2007), 351n48.

20. A concrete point of reference here is the study by Annemarie Mol and John Law, "Regions, Networks, and Fluids: Anaemia and Social Topology," *Social Studies of Science* 24, no. 4 (1994): 641–672.

21. Gilles Deleuze and Félix Guattari, *Anti-Oedipus: Capitalism and Schizophrenia*, trans. Robert Hurley, Mark Seem, and Helen R. Lane (Minneapolis: University of Minnesota Press, 1992), 9–10.

22. Gilbert Simondon, *Du mode d'existence des objets techniques* [1958], 3rd ed. (Paris: Aubier, 1989), 60.

23. See http://modesofexistence.org. For background information concerning the *Inquiry*'s status within the body of Latour's work, see Bruno Latour, "Biographie d'une enquête: À propos d'un livre sur les modes d'existence," *Archives de Philosophie* 75 (2012): 549–566.

24. Latour, "Coming Out as a Philosopher."

25. Latour, "Reflections on Etienne Souriau's *Les différents modes d'existence*," 309.

26. John L. Austin, *How to Do Things with Words*, 2nd ed. (Oxford: Oxford University Press, 1978).

27. Gilles Deleuze and Félix Guattari, *What Is Philosophy?*, trans. Hugh Tomlinson and Graham Burchell (New York: Columbia University Press, 1994), 35–60.

28. See Latour, "Reflections on Etienne Souriau's *Les différents modes d'existence*," 307–311. See also Bruno Latour and Isabelle Stengers, "Le sphinx de l'œuvre," in Etienne Souriau, *Les différents modes d'existence*, 2nd ed. (Paris: Presses universitaires de France, 2009), 12.

29. Simondon, *Du mode d'existence des objets techniques.*

30. Etienne Souriau, ed., *L'univers filmique* (Paris: Flammarion, 1953).

31. Latour, "Reflections on Etienne Souriau's *Les différents modes d'existence*," 317, 323.

32. Mikel Dufrenne, *Phénoménologie de l'expérience esthétique*, 2 vols. (Paris: Presses universitaires de France, 1953).

33. Etienne Souriau, *Les différents modes d'existence* (Paris: Presses universitaires de France, 1943), 64.

34. Oskar Becker, *Mathematische Existenz. Untersuchungen zur Logik und Ontologie mathematischer Phänomene* [1927], 2nd ed. (Tübingen: Niemeyer, 1973), 181. On Becker, see Annemarie Gethmann-Siefert, "A Phenomenological Aesthetics: Oskar Becker's Coupling of Epistemology and Ontology," *New Yearbook for Phenomenology and Phenomenological Philosophy* 2 (2002): 137–177.

35. Roman Ingarden, *The Literary Work of Art: An Investigation on the Borderlines of Ontology, Logic, and Theory of Literature*, trans. George G.

Grabowicz (Evanston, Ill.: Northwestern University Press, 1973). The German version was published in 1927.

36. Martin Heidegger, "Letter on 'Humanism'" [1949], in *Pathmarks*, ed. William McNeill (Cambridge: Cambridge University Press, 1998), 239–276, where Heidegger speaks of the "homelessness" of modern life (257–262). See also Peter Sloterdijk, *Regeln für den Menschenpark: Ein Antwortschreiben zu Heideggers Brief über den Humanismus* (Frankfurt: Suhrkamp, 1999).

37. These stories are clearly marked in the French version of the book. See Bruno Latour, *Enquête sur les modes d'existence: Une anthropologie des Modernes* (Paris: La Découverte, 2012), 484–485.

38. Robert Michael Brain, "The Ontology of the Questionnaire: Max Weber on Measurement and Mass Investigation," *Studies in History and Philosophy of Science* 32, no. 4 (2001): 647–684; Christophe Prochasson, "L'enquêteur, le savant et le démocrate: Les significations politique de l'enquête," *Mille neuf cent: Revue d'histoire intellectuelle* 22, no. 1 (2004): 7–14.

CONCLUSION

1. John H. Zammito, *A Nice Derangement of Epistemes: Postpositivism in the Study of Science from Quine to Latour* (Chicago: University of Chicago Press, 2004), 152.

2. Hermann Melville, "Bartleby, the Scrivener; A Story of Wall-Street," in *The Piazza Tales and Other Prose Pieces, 1839–1860* (Evanston, Ill.: Northwestern University Press, 1987), 20.

3. Isabelle Stengers, "The Cosmopolitical Proposal," in *Making Things Public: Atmospheres of Democracy*, ed. Bruno Latour and Peter Weibel (Cambridge, Mass.: The MIT Press, 2005), 994–1003; Gilles Deleuze and Félix Guattari, *What Is Philosophy?*, trans. Hugh Tomlinson and Graham Burchell (New York: Columbia University Press, 1994), 61–64.

4. Stengers, "The Cosmopolitical Proposal," 995. See also MoE 480: "Well, yes, you have to be *idiotic* to throw yourself into something like this [i.e., the *Inquiry*]."

5. Peter Sloterdijk, "Excursus 7. On the Difference Between an Idiot and an Angel," in *Spheres*, vol. 1, *Bubbles: Microspherology*, trans. Wieland Hoban (Los Angeles: Semiotext(e), 2011), 473.

6. Martin Heidegger, *Sein und Zeit*, 17th ed. (Tübingen: Niemeyer, 1993), 78. See also Heidegger, *Being and Time: A Translation of* Sein und Zeit, trans. Joan Stambaugh (Albany, N.Y.: State University of New York Press, 1996), 73, where the remarkable term "*Zeigzeug*" is rendered as "pointer."

7. Charles Péguy, "Note sur M. Bergson et la philosophie bergsonienne," in *Œuvres en prose, 1909–1914* (Paris: Gallimard, 1961), 1551–1552.

8. See the cover of the paperback edition, Bruno Latour, *Nous n'avons jamais été modernes: Essai d'anthropologie symmétrique* (Paris: La Découverte, 2006).

9. Bruno Latour, "What Is Iconoclash? Or Is There a World Beyond the Image Wars?" in *Iconoclash: Beyond the Image Wars in Science, Religion, and Art*, ed. Bruno Latour and Peter Weibel (Karlsruhe: ZKM Center for Art and Media/ Cambridge, Mass.: The MIT Press, 2002), 34.

10. See, for example, Oskar Becker, *Mathematische Existenz: Untersuchungen zur Logik und Ontologie mathematischer Phänomene* [1927], 2nd ed. (Tübingen: Niemeyer, 1973); Roman Ingarden, *The Literary Work of Art: An Investigation on the Borderlines of Ontology, Logic, and Theory of Literature* [1927], trans. George G. Grabowicz (Evanston, Ill.: Northwestern University Press, 1973).

11. Georges Canguilhem, "On the History of the Life Sciences Since Darwin," in *Ideology and Rationality in the History of the Life Sciences* [1977], trans. Arthur Goldhammer (Cambridge, Mass.: The MIT Press, 1988), 117; Ian Hacking, *Representing and Intervening: Introductory Topics in the Philosophy of Natural Science* (Cambridge: Cambridge University Press, 1983), 220–232.

12. Hans-Jörg Rheinberger, *Toward a History of Epistemic Things: Synthesizing Proteins in the Test Tube* (Stanford, Calif.: Stanford University Press, 1997), 110–113; Friedrich A. Kittler, *Discourse Networks 1800/1900*, trans. Michael Metteer, with Chris Cullens (Stanford, Calif.: Stanford University Press, 1990), 369–372.

13. Peter Galison, *Einstein's Clocks, Poincaré's Maps: Empires of Time* (New York: Norton, 2003); Simon Schaffer, "Late Victorian Metrology and Its Instrumentation: A Manufactory of Ohms," in *Invisible Connections: Instruments, Institutions, and Science*, ed. Robert Bud and Susan E. Cozzens (Bellingham, Wash.: SPIE Optical Engineering Press, 1992), 25–56; Myles Jackson, *Spectrum of Belief: Joseph von Fraunhofer and the Craft of Precision Optics* (Cambridge, Mass.: The MIT Press, 2000); Hannah Landecker, *Culturing Life: How Cells Became Technologies* (Cambridge, Mass.: Harvard University Press, 2006); Cornelia Vismann, *Files: Law and Media Technology*, trans. Geoffrey Winthrop-Young (Stanford, Calif.: Stanford University Press, 2008).

14. Robert Crease, Don Ihde, Casper Bruun Jensen, and Evan Selinger, "Interview with Bruno Latour," in *Chasing Technoscience: Matrix for Reality*, ed. Don Ihde and Evan Selinger (Bloomington: Indiana University Press, 2003), 19.

BIBLIOGRAPHY

A complete list of Bruno Latour's publications, which is constantly updated, as well as other resources (texts, videos, audio files) can be found on Latour's personal website: http://www.bruno-latour.fr/.

For additional literature on Latour and actor-network theory, see John Law's annotated bibliography: http://www.lancs.ac.uk/fass/centres/css/ant/antres.htm.

SELECTED PUBLICATIONS ON LATOUR

Blok, Anders, and Torben Elgaard Jensen. *Bruno Latour: Hybrid Thoughts in a Hybrid World.* London: Routledge, 2011.

Golinski, Jan. "Science and Religion in Postmodern Perspective: The Case of Bruno Latour." In *Science and Religion: New Historical Perspectives*, ed. Thomas Dixon, Geoffrey Cantor, and Stephen Pumfrey, 50–68. Cambridge: Cambridge University Press, 2010.

Harman, Graham. *Prince of Networks: Bruno Latour and Metaphysics.* Melbourne: Re:Press, 2009.

Keuchyean, Razmig. "L'imagination constructiviste: Une enquête au *Centre de Sociologie de l'Innovation.*" *L'année sociologique* 58, no. 2 (2008): 409–434.

Kneer, Georg, Markus Schroer, and Erhard Schüttpelz, eds. *Bruno Latours Kollektive: Kontroversen zur Entgrenzung des Sozialen.* Frankfurt: Suhrkamp, 2008.

Kochan, Jeff. "Latour's Heidegger." *Social Studies of Science* 40, no. 4 (2010): 579–598.

Law, John, and John Hassard, eds. *Actor Network Theory and After.* Oxford: Blackwell/Malden, Mass.: The Sociological Review, 1999.

Zammito, John R. *A Nice Derangement of Epistemes: Postpositivism in the Study of Science from Quine to Latour.* Chicago: University of Chicago Press, 2004.

FURTHER READING

Czarniawska, Barbara. *A Theory of Organizing*. Cheltenham: Elgar, 2008.
Douglas, Mary. *Purity and Danger: An Analysis of Concepts of Pollution and Taboo*. London: Routledge & Kegan Paul, 1966.
Galison, Peter. *Einstein's Clocks, Poincaré's Maps: Empires of Time*. New York: Norton, 2003.
Jackson, Myles. *Spectrum of Belief: Joseph von Fraunhofer and the Craft of Precision Optics*. Cambridge, Mass.: The MIT Press, 2000.
Joerges, Bernward. *Technik—Körper der Gesellschaft. Arbeiten zur Techniksoziologie*. Frankfurt: Suhrkamp, 1996.
Robinson, Thomas M., ed. *Contrasting Arguments: An Edition of the* Dissoi Logoi. New York: Arno, 1979.
Schaffer, Simon. "Late Victorian Metrology and Its Instrumentation: A Manufactory of Ohms." In *Invisible Connections: Instruments, Institutions, and Science*, ed. Robert Bud and Susan E. Cozzens, 25–56. Bellingham: SPIE Optical Engineering Press, 1992.
Schillmeier, Michael. *Rethinking Disability: Bodies, Senses, and Things*. London: Routledge, 2010.
Vismann, Cornelia. *Files: Law and Media Technology*. Trans. Geoffrey Winthrop-Young. Stanford, Calif.: Stanford University Press, 2008.
Yaneva, Albena. "Scaling Up and Down: Extraction Trials in Architectural Design." *Social Studies of Science* 35, no. 6 (2005): 867–894.

INDEX

Stefanos Geroulanos and Todd Meyers, *series editors*

Georges Canguilhem, *Knowledge of Life.* Translated by Stefanos Geroulanos and Daniela Ginsburg, Introduction by Paola Marrati and Todd Meyers.

Henri Atlan, *Selected Writings: On Self-Organization, Philosophy, Bioethics, and Judaism.* Edited and with an Introduction by Stefanos Geroulanos and Todd Meyers.

Catherine Malabou, *The New Wounded: From Neurosis to Brain Damage.* Translated by Steven Miller.

François Delaporte, *Chagas Disease: History of a Continent's Scourge.* Translated by Arthur Goldhammer, Foreword by Todd Meyers.

Jonathan Strauss, *Human Remains: Medicine, Death, and Desire in Nineteenth-Century Paris.*

Georges Canguilhem, *Writings on Medicine.* Translated and with an Introduction by Stefanos Geroulanos and Todd Meyers.

François Delaporte, *Figures of Medicine: Blood, Face Transplants, Parasites.* Translated by Nils F. Schott, Foreword by Christopher Lawrence.

Juan Manuel Garrido, *On Time, Being, and Hunger: Challenging the Traditional Way of Thinking Life.*

Pamela Reynolds, *War in Worcester: Youth and the Apartheid State.*

Vanessa Lemm and Miguel Vatter, eds., *The Government of Life: Foucault, Biopolitics, and Neoliberalism.*

Henning Schmidgen, *The Helmholtz Curves: Tracing Lost Time.* Translated by Nils F. Schott.

Henning Schmidgen, *Bruno Latour in Pieces: An Intellectual Biography.* Translated by Gloria Custance.

Veena Das, *Affliction: Health, Disease, Poverty.*

Kathleen Frederickson, *The Ploy of Instinct: Victorian Sciences of Nature and Sexuality in Liberal Governance.*

Roma Chatterji (ed.), *Wording the World: Veena Das and Scenes of Inheritance.*

Jean-Luc Nancy and Aurélien Barrau, *What's These Worlds Coming To?* Translated by Travis Holloway and Flor Méchain. Foreword by David Pettigrew.